JAVA SERVER PAGES

JAVA SERVER PAGES

Ankur Saxena

Assistant Professor
Amity Institute of Biotechnology
Amity University
Noida
Uttar Pradesh

MJP PUBLISHERS

CHENNAI NEW DELHI TIRUNELVELI

Cataloguing-in-Publication Data

Saxena, Ankur (1979 -).
Java server pages / by Ankur Saxena.-
Chennai: MJP Publishers, 2012.
xiv, 450 p.; 24 cm.
Includes References.
ISBN 978-81-8094-133-7 (pb.)
1. Computer Science, Java I. Title
005.262 SAX MJP 139

ISBN 978-81-8094-133-7

Printed and bound in India

MJP PUBLISHERS
New No. 5, Muthu Kalathy Street
Triplicane
Chennai 600 005

Publisher : J.C. Pillai
Managing Editor : C. Sajeesh Kumar
Project Editor : P. Parvath Radha
Acquisitions Editor : C. Janarthanan
Editorial Team : B. Ramalakshmi, V.R. Padma, R. Hemalatha, B. Annalakshmi, S. Jeevasruthi, N. Yamuna Devi, C. Devi
CIP Data : Prof. K. Hariharan, Librarian
RKM Vivekananda College, Chennai.

ACKNOWLEDGEMENTS

Thank you Lord, for everything. I never have words to say to you what I want to say. But, thank you very much.

Many thanks to all my students, who during the classes brought up many issues and gave constructive feedback.

Heartiest thanks to my corporate friends in various parts of the country, who keep sending suggestions and feedback about the content, presentation and design of the book.

I am thankful to Mr. Sooraj goel for patiently waiting for this book.

This book would not be in its present shape without the dedicated efforts of Smitha M.S. Thanks to this wonderful lady who makes wonderful book happen. I also owe thanks to other member of my team: Dr. Nidhi Chaudhary, Mr. Ankur Chaurasia, Mr. Praveen for giving ample time to this book's design .

Thanks to Mr Ankit Saxena who took extra pains for the production of this book.

Finally, thanks to my family members and friends for their support. I owe heartiest thanks to my parents, who are the foundation stone of my book.

I am grateful to MJP publishers for bringing out this book.

Ankur Saxena

PREFACE

The book "JAVA SERVER PAGES" for engineering, MCA and software professionals has been prepared keeping in mind the requirements of professionals as well as those who deal with this technology. The main thrust of the book is to deal with advance features of JSP.

Starting with an introductory chapter, the book covers all the features of java server pages required for software professionals as well as engineering students. An easy to learn approach has been adopted throughout. Special attention has been paid to make the material comprehensive, concise and practical. Simple and lucid language has been used to avoid any confusion while using this technology.

The concepts in this book have been presented with suitable examples whenever required. This enables the reader to use the book with confidence for self study. Each chapter contains a number of review questions which are carefully designed to provide an insight into the subject. Efforts have been made to avoid errors specially programming errors. The programming examples have been tested on tomcat (Application server), latest version.

Any suggestion for improvement of the book will be gratefully acknowledged.

Ankur Saxena

CONTENTS

1
INTRODUCTION TO JSP

In today's environment, most of the server side websites are based on user request over dynamic pages, which are capable of communicating with the databases working behind the scene. Database is a very convenient way to store the data of users and other things. Java Database Connection (JDBC) provides excellent database connectivity in a heterogeneous database environment. Using Java Server Pages (JSP) and JDBC, it is very easy to develop database-driven web applications.

Nowadays, Java is recognized as a suitable and secured language for the server side programming. Companies dealing with finance and other related fields are now quickly recognizing the potential and security of web applications designed with the help of Java. Various companies use different operating systems or platforms as per the requirement of their system and the key feature of platform independence makes Java more reliable and efficient over other languages.

1.1 HISTORY OF JSP

JSP, a specification of Sun Microsystems, first appeared in 1998. The official versions, 1.0 and 1.1, both appeared in 1999, and both were very popular. Later on, version 1.2 appeared in 2001, and is the most popular implementation yet; this is the version we will be using in this book. Internally, each JSP page is actually converted into a servlet. Servlets are server side components that provide a powerful mechanism for developing server side programs by the server.

A JSP page is a text-based document that describes how to process a request to create a response. JSP is a Java-based technology that simplifies the process of developing dynamic websites. With JSP, web designers and developers can quickly incorporate dynamic elements into web pages using embedded Java and simple markup tags. These tags provide the HTML (Hypertext Markup Language) designer with a way to access data and business logic stored inside Java object. Java Server Pages are text files with the extension .jsp, which takes the place of traditional HTML pages. JSP technology enables you to mix regular, static HTML with dynamically generated content from servlet. You simply write the regular HTML in the normal manner, using familiar web-page-building tools. You then enclose the code for the dynamic parts in special tags. JSP provides excellent server-side scripting support for creating database-driven web applications. JSP enables the developers to directly insert Java code into .jsp file. This makes the development process very simple and its maintenance also becomes very easy.

1.2 WHY USE JSP

JSP technology allows web developers and designers to quickly develop and easily maintain, information-rich, fast and dynamic web pages that leverage existing business systems. As a part of the Java technology family, JSP technology allows rapid development of web-based applications that are platform-independent, which means, "develop once and run anywhere". JSP technology seperates the user interface from content generation, enabling designers to change the overall page layout without altering the underlying dynamic content. Java Server Pages inherits the concepts of applications, ServletContexts, Sessions, Requests, and Responses from the Java Servlets Application Programming Interface (API) and offer the same portability, performance, and scalability as servlets.

1.2.1 Benefits for Developers

If you are a web page developer or designer who is familiar with HTML, you can:

Use JSP without having to learn the Java language You can use JSP technology without learning how to write Java scriptlets. Although scriptlets are no longer required to generate dynamic content, they are still supported to provide backward compatibility.

Extend the JSP language Java tag library developers and designers can extend the JSP language with "simple tag handlers," which utilize a new, much simpler and cleaner, tag extension API. This spurs the growing number of pluggable, reusable tag libraries available, which in turn reduces the amount of code needed to write powerful web applications.

Easily write and maintain pages The Java Server Pages Standard Tag Library (JSTL) expression language is now integrated into JSP technology and has been upgraded to support functions. The expression language can now be used instead of scriptlet expressions.

JSP technology uses tags similar to XML that encapsulates, the logic that generates the content for the page. The application logic can reside in server-based resources (such as JavaBeans component architecture) that the page accesses with these tags. By separating the page logic from its design and display which supports a reusable component-based design, JSP technology makes it faster and easier than ever to build web-based applications. JSP allows Java code and certain pre-defined actions to be interleaved with static web markup content, with the resulting page being compiled and executed on the server to deliver an HTML or XML document. The compiled pages and any Java-dependent libraries use Java bytecode rather than a native software format, and must therefore be executed within a Java virtual machine (JVM) that integrates with the host operating system to provide an abstract platform-neutral environment.

1.3 THE JSP API

The JSP technology is based on the JSP API (Application Programming Interface) that consists of two packages:

1. javax.servlet.jsp
2. javax.servlet.jsp.tagext

Along with these two packages, JSP also needs the two servlet packages, i.e., javax.servlet and javax.servlet.http.

The javax.servlet.jsp package has two interfaces and four classes which are as follows.

The two interfaces are

1. JspPage
2. HttpJspPage

The four classes are:

1. JspEngineInfo
2. JspFactory
3. JspWriter
4. PageContext

In addition, there are also two exception classes: JspException and JspError.

1.3.1 The JspPage Interface

The JspPage interface describes the generic interaction that a JSP Page Implementation class must satisfy. Pages that use the HTTP protocol are described by the HttpJspPage interface. The interface defines a protocol with three methods. Only two of them: jspInit() and jspDestroy() are part of this interface as the signature of the third method jspService() depends on the specific protocol used and cannot be expressed in a generic way in Java. A class implementing this interface is responsible for invoking the above methods at the appropriate time, based on the corresponding servlet-based method invocations. The jspInit() and jspDestroy() methods can be defined by a JSP author, but the jspService() method is defined automatically by the JSP processor based on the contents of the JSP page.

- jspService() method corresponds to the body of the JSP page. This method is defined automatically by the JSP container and should never be defined by the JSP page. If a super class is specified using the extends attribute, that super class may choose to perform some actions in its service() method before or after calling the jspservice() method.
- The jspInit() method is invoked when the JSP page is initialized. It is the responsibility of the JSP implementation (and of the class mentioned by

the extends attribute, if present) that at this point invocations to the getServletConfig() method will return the desired value. A JSP page can override this method by including a definition for it in a declaration element. A JSP page should re-define the init() method from servlet.

- The jspDestroy() method is invoked when the JSP page is about to be destroyed. A JSP page can override this method by including a definition for it in a declaration element. A JSP page should re-define the destroy() method from servlet.

1.3.2 The HttpJspPage Interface

This interface directly extends the JspPage interface. There is only one method, jspService. This method is called by the JSP container to generate the content of the JSP page. The jspservice has the following signature.

Public void _jspService (HttpServletRequest request, HttpServletResponse response) throws ServletException, IOException.

1.3.3 The JspEngineInfo Interface

The JspEngineInfo class is an abstract class that provides information on the JSP container. Only one method, getSpecificationVersion, returns the JSP container's version number. Because this is the only method currently available, this class does not have much use. You can obtain a JspEngineInfo object using the getEngineInfo method of the JspFactory class.

1.3.4 The JspFactory Class

The JspFactory class is an abstract class that provides methods for obtaining other objects needed for the JSP page processing. The class has the static method, getDefaultFactory, that returns a JspFactory object, from which a PageContext and a JspEngineInfo object can be obtained that are useful for the JSP page processing.

1.3.5 The JspWriter Class

The JspWriter class is derived from the java.io.Writer class and represents a Writer that can be used to write to the client browser. Of its many methods, the most important are the println and print methods. Both provide enough overloads that ensure that you can write any page of data. The difference between print and println is that println always adds the new line character to the printed data.

Additional method allows you to manipulate the buffer. For instance, the clear method clears the buffer. It throws an exception if some of the buffer's content has already been flushed. Similar to clear is the clearBuffer method, which clears the buffer but never throws any exception if any of the buffer's contents have been flushed.

1.3.6 The PageContext Class

PageContext represents a class that provides methods that are implementation-dependent. The PageContext class itself is abstract, so in the jspService method of a JSP servlet class, a PageContext object is obtained by calling the getPageContext method of the JspFactory class. The PageContext class provides methods that are used to create other objects. Other methods that return servlet-related objects include the following:

- getRequest returns a ServletRequest object
- getResponse returns a ServletResponse object
- getServletConfig returns a ServletConfig object
- getServletContext returns a ServletContext object
- getSession returns an HttpSession object

1.4 DYNAMIC CONTENT TECHNOLOGIES

For the simplest web requests, a browser requests an HTML document, and the web server finds the corresponding file and returns it. But in dynamic web content, the web server does some additional processing of the corresponding request in order to generate a customized response. In addition to the URL of the request, the form of this customized response might be based on the date and time, the location on the network from which the request was made, or on some representation of the identity of the user making the request. Indeed, the exact details of the response might depend upon a combination of some or all of these factors. Some of the technologies prevailing are discussed here to give an idea about them.

1.4.1 Common Gateway Interface (CGI)

The first standard for dynamic web content was based on the Common Gateway Interface or CGI, which specifies a mechanism for web servers to generate response at runtime. The Perl language is a popular choice for writing CGI programs, but CGI codes can be written in any language that can be called as a standalone program by HTTP server. For example, a CGI program could be written in any scripting language supported by the local operating system. Alternately, it could be written in C and compiled into native object code. CGI programs could even be written as Java Applications.

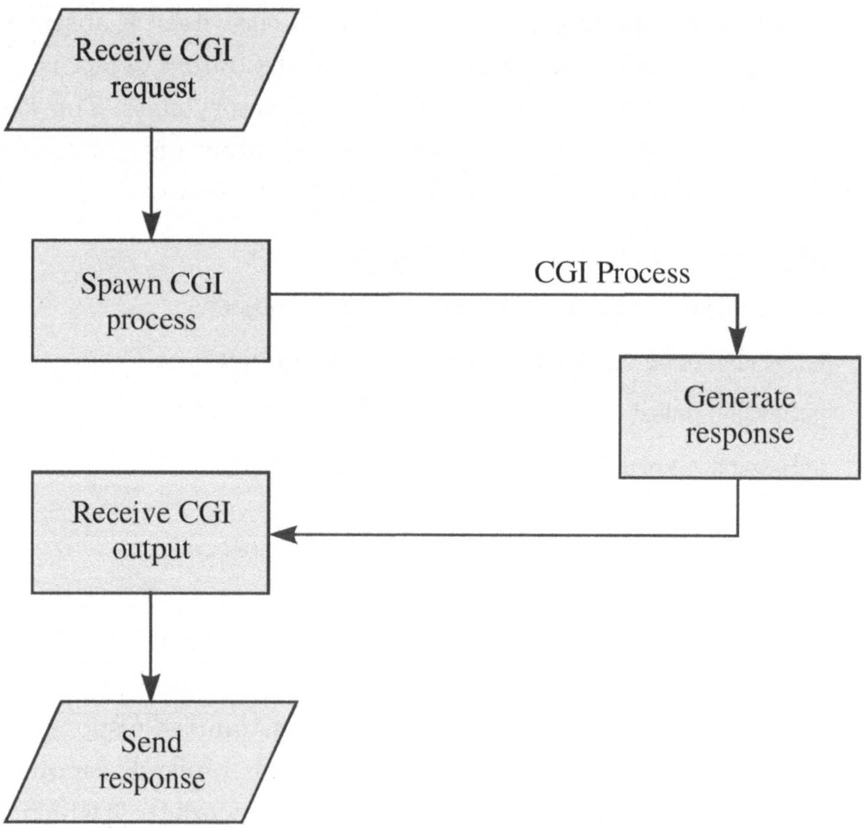

Figure 1.1 Server process for running CGI programs

As indicated in Figure 1.1, CGI programs run outside the web server. This means that a new process must be started to execute a CGI program. There is some overhead associated with creating and communicating with this separate process and each process requires its own share of the local machine's memory recourses. Furthermore, CGI programs are designed to handle only a single request, after which they return their results to the web server and exit. This means that each time a request for dynamic content is received by the web browser, it must start a new process for running the corresponding CGI program for that specific request, send the requested information, wait for the results, then pass those results back as a response to the browser. These days, it is common for popular websites to be handling thousands of simultaneous requests, so even if the inefficiencies of individual CGI program execution are fairly low, they quickly add up to significant performance penalties.

1.4.2 ColdFusion

ColdFusion provides a set of HTML-like tags which were initially targeted at embedding database queries into web pages, but it has since been extended to support a wide variety of data source for dynamic content generation. It has a single consistent style of syntax throughout the page. ColdFusion supports both UNIX and Microsoft Windows platforms.

1.5 JSP VERSUS ASP

Microsoft tried to fill this hole with Active Server Pages (ASP), which allows developers to use simple scripting to access the server and its extensions. But, while ASP gives you an efficient way to return dynamic content, it essentially limits you to Microsoft platforms, and even the simplest of scripting mistakes can cause the server to crash or hang, effectively bringing down your website. Apparently in response to ASP, Sun Microsystems gave JavaServer Pages (JSP) technology, which is based entirely upon Sun's popular Java programming language and gives developers the advantages of developing Java in a more relaxed, script-like environment. Behind the scenes, JSP pages are dynamically assembled into Servlets, which are simply Java classes. This means JSP supports the same modularity, reusability, platform-independence, and access to Java APIs that Java programming supports.

JSP also bests ASP by supporting the definition of tags that abstract functionality within a page, so tags can be defined in a tag library and then used within any JSP page. This makes a better separation of page content from its code, which is one of web development's prime directives. The lesser the code scattered throughout the pages of a website, the easier the site is to maintain. Global changes need to be made to the tags defined in a central library, making time-consuming, page-by-page fix things of the past. For XML users, good news is that JSP pages can be written in well-formed, valid XML by using XSLT. (For those of you who are not XML-spawn, this simply means that JSP can make full use of a very powerful and widely accepted data format and all the tools and support that go with it.). And unlike ASP, JSP is lot less platform-specific and it does not rely heavily on the company that created it for support or performance improvements. That shows, there are some similarities between JSP and ASP. Here, let us take a closer look at how both are compared.

Active Server Pages (ASP) is the Microsoft's solution for providing dynamic Web content. Actually, ASP looks very similar to JSP; both use custom tags to implement business logic and text (HTML) for invariant web page parts. However, the drawback is in the details, as described in the following.

1. ASP uses VBScript or JScript, a Microsoft flavour of JavaScript, as its scripting language, whereas JSP uses Java, a more powerful language than VBScript or JScript.
2. The ASP developer typically uses a Microsoft's web server platform or requires a third party product that permits ASP execution on non-Microsoft platforms, while JSP developer has a wide variety of web server platforms available for use.

Note that, these third parties must port Microsoft software components, such as ActiveX, to different platforms in order for ASP to be used on these platforms. An ASP is interpreted every time the page is invoked, whereas a JSP is interpreted only the first time the page is invoked (or when the page is changed). The different features of JSP and ASP technology are given in Table 1.1.

Table 1.1 Comparison of JSP and ASP technology

JSP technology	ASP technology	Definition
Any web server, including Apache Tomcat, Weblogic, Jboss, IIS	Microsoft IIS or Personal web server	Supporting web server
Most popular platforms, including the Solaris Operation System, Microsoft Windows, Mac OS, Linux and Other UNIX platform implementations	Microsoft Windows	Platform Support
Request	Request	Access to request information. This includes such items as input/form parameters, HTTP request header information, etc.
Response	Response	Access to response information and construction. This includes output MIME type and other HTTP response headers as well as the output stream itself.
Session	Session	Access to session information.
Application	Application	Access to application information.
Out	Response.write	Access to the response output stream. JSP provides a convenience object for direct output to this stream, whereas ASP uses Response object method.

The different coding examples of JSP and ASP technology are given in Table 1.2.

Table 1.2 Coding examples

JSP	ASP

<%@ page language="java" %> <TITLE>
Simple JSP Code</TITLE> Anyone can count
like this:
 	
<%
for (int count = 1; count < 6; count++)
{
out.println(count + "
");
}
count = 1000000;
%>
It would take a long time, however, to count
to <%= count %>.
	<%@ page language="java" %>
	<TITLE>Simple ASP Code</TITLE>
	Anyone can count like this:
	
	<%
	for (int count = 1; count < 6; count++)
	{
	Response.Write(count + "
	");
	}
	count = 1000000;
	%>
	It would take a long time, however, to
	count to <%= count %>.

1.6 FEATURES OF JSP

JSP or Java Server Pages was developed by Sun Microsystems. JSP technology is object-oriented programming language and is based on Java language.

Some of its salient features are as follows:

- JSP is widely used for developing dynamic websites.
- JSP is used for creating database driven web applications because it provides superior server-side scripting support.

Some of the reasons why JSP is most popular are as follows:

Simplifies the process of development JSP pages are installed simply as web pages using natural structure of web server document tree. It allows programmers to insert the Java code directly into the JSP file, making the development process easier. Although JSP files are HTML files, they use special tags containing the Java source code which provides a dynamic ability. JSP uses a combination of tags and scripting to create dynamic web pages.

Portability The Java feature of "write once, run anywhere" is applicable to JSP. JSP is also platform-independent, making it portable across any platform and therefore, multi-platform. In other words, JSP runs on virtually any environment that supports JAVA servlet and hence it is compatible. It is possible for the programmer to take a JSP file and move it to another platform, JSP servlet engine or web server.

Efficiency A language for developing JSP pages, which are text-based documents that describe how to process a request and construct a response. JSP uses pure java and takes the advantage of its object-oriented nature. As soon as the request is received, the JSP page gets loaded into the web server's memory and then transactions are carried out within a minimal time period, handling JSP pages with efficiency. JSP is constructed for accessing server-side objects. JSP uses a combination of tags and scripting to create dynamic web pages.

Reusability The JSP page uses the components like EJB and JavaBeans which are reusable. This gives the JSP reusability capabilities.

Robustness JSP offers a robust platform for web development.

Independency of layers There is a clear separation between presentation and implementation layers. The HTML on the web browser of the client is displayed as a presentation layer. The JSP on the server is displayed in the implementation layer. Programmers who want to work with HTML can work independently without the help of the Java developers, but those working in the implementation layer have to be dependent on Java developers. Java Server Pages (JSP) is the way to separate the look of the web page from the corresponding content.

Integration of JSP with other sources like JDBC, servlet and so on The combination of JDBC and JSP works very well. JSP is used for generating dynamic web pages. It is essential that data in these systems are maintained efficiently and securely. The programmer must have a good database and database connectivity. This is achieved by JDBC through its excellent database connectivity in heterogeneous database system. This database system is used for the integration of heterogeneous database management systems in presenting a single-query-interface system.

Simplification of process The JSP language has a simple development and maintenance process. A JSP file that has the extension .jsp is converted into a servlet.java which is dynamically compiled, loaded and executed. Only when there is a change in a JSP file, the conversion, compilation, and loading process is then performed.

1.7 GENERAL RULES OF SYNTAX

Some of the general rules of syntax are as follows:

Java is a case-sensitive language Being a part of Java language, JSP is also a case- sensitive language.

Example

Incorrect code	Correct code

<%@ page language="java" %>
<TITLE>Simple JSP Code</TITLE>
Anyone can count like this:
 	
<%
for (int count = 1; count < 6; count++)
{
Out.println(count + "
"); **//Error**
}
count = 1000000;
%>
	<%@ page language="java" %>
	<TITLE>Simple JSP Code</TITLE>
	Anyone can count like this:
	
	<%
	for (int count = 1; count < 6; count++)
	{
	out.println(count + "
	");
	}
	count = 1000000;
	%>

Avoid duplicate local JSP variable names Avoid duplicate local variable names in JSP. This can happen when other JSP's are included.

Example

File1.jsp	File2.jsp
<%@ include file="File2.jsp" %> <% int var = 0; **// VIOLATION** %>	<% int var = 0; %>

Variable name used is JSP programs should not include any JSP keyword, space or special characters except an underscore (_).

Example

Incorrect code	**Correct code**

<%@ page language="java" %>
<TITLE>Simple JSP Code</TITLE>
Anyone can count like this:
 	
<%
for (int col num = 1; col num < 6; col num++) **//Error**
{
out.println(col num + "
"); **//Error**
}
%>
	<%@ page language="java" %>
	<TITLE>Simple JSP Code</TITLE>
	Anyone can count like this:
	
	<%
	for (int col_num = 1; col_num < 6; col_num++)
	{
	out.println(col_num + "
	");
	}
	%>

Cyclic inclusion of JSP's Detects the infinite loop caused by including multiple JSPs.

Example

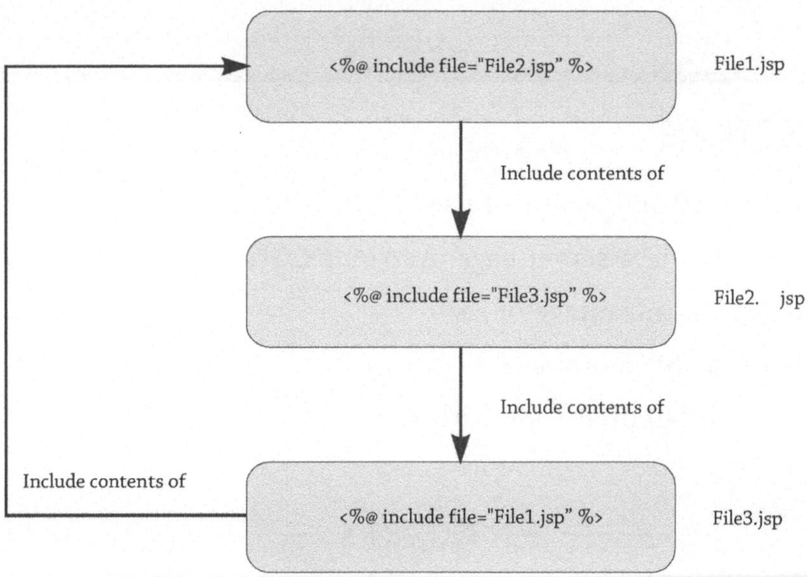

Never use reserved JSP local variable names Never use local variable names used by JSP Servlets, Beans, etc. as this can result in duplicate variable names.

Example

```
File1.jsp

<%

int request=-1;      //VIOLATION: Request is a reserved keyword in JSP
```

Always provide JSP header comments Complete JSP header comments should always be included within the file.

Example

Sample code	Should be written as
<html> <head> <title>Notice and Result</title> </head> <body topmargin=0> .... </body> </html>	<%-- This JSP displays the login form -%> <html> <head> <title>Notice and Result</title> </head> <body topmargin=0> .... </body> </html>

REVIEW QUESTIONS

1. What is JSP and what is it used for?
2. How does a java server page overcome CGI?
3. What is the advantage of JSP?
4. What is a JSP container?
5. How to differentiate JSP with ASP.

2

JSP COMPARISON WITH SERVLET (ON THE BASIS OF LIFE CYCLE)

2..1 JSP LIFE CYCLE

JSP never outputs the content directly to the browser. Instead, JSP relies on initial server-side processing process which translates JSP page into the JSP Page class. The page class will then handle all the requests made of JSP. A JSP page is first converted into a servlet and then that servlet gets compiled and executed like any other servlet. The translation of a JSP page into the corresponding servlet is done by the JSP Engine of the underlying Web Container. The JSP life cycle is described in Figures 2.1 and 2.2.

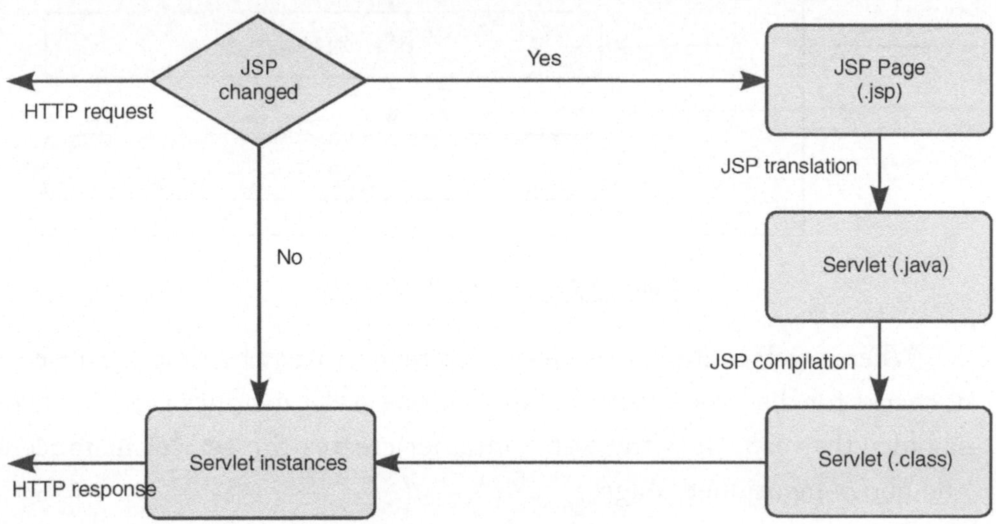

Figure 2.1 JSP lifecycle

The entire life cycle of a typical JSP page can be summarized in the following phases.

- Translation
- Compilation
- Class loading
- Initialization
- Execution
- Service
- Finalization
- Destroy

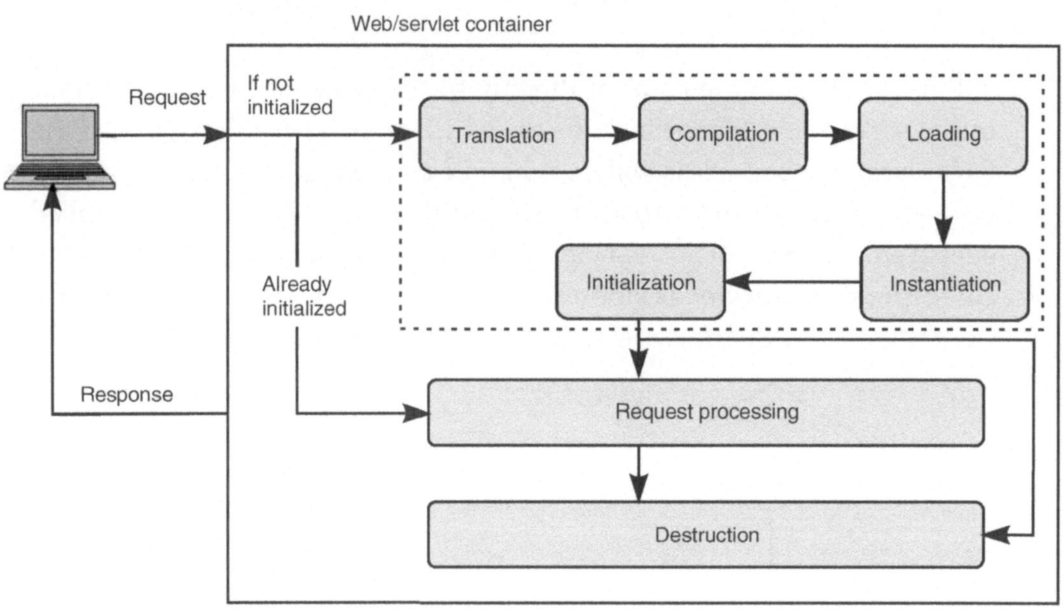

Figure 2.2 Complete JSP life cycle

When a web container receives a jsp request (may be first or subsequent), it checks for the jsp's servlet instance. If no servlet instance is available or if it is older than jsp, then the web container creates the servlet instance using the above-mentioned stages.

2.1.1 JSP Page Translation

A java servlet file is generated from the JSP source file. This is the first step in its tedious multiple-phase life cycle. In the translation phase, the container validates the syntactic correctness of the JSP pages and tag files. The container interprets the standard directives and actions, and the custom actions referencing tag libraries used in the page. In the translation phase, JSP engine checks for the JSP syntax and translates the JSP page into its page implementation class if the syntax is correct. This class is actually a standard Java servlet. After that, JSP engine compiles the source file into class file and is ready for use. If the container receives the request, it checks for the changes of the JSP page since it was last translated. If no changes were made, it just loads the servlet otherwise the process of check, translate and compile occurs again. Because the compilation process takes time, JSP engine minimizes the time to increase the performance of the page processing. In this phase the JSP page is translated into the corresponding servlet. This translated servlet is different from a normal servlet only in the sense that it implements the interface javax.servlet.jsp.HttpJspPage. Well if you look at this interface closely, you will easily notice that a translated JSP page implements this interface just to have the JSP specific features. This HttpJspPage is just a wrapper of the Servlet interface as HttpJspPage interface extends the JspPage interface and the JspPage interface extends the javax.servlet.Servlet interface. Thus, we see that the translated Servlet also internally implements the javax.servlet. Servlet interface (otherwise we would not have called it a servlet). In addition it implements few more interfaces to support the JSP technology. Web container translates (converts) the jsp code into a servlet code. This means that jsp is actually a servlet. After this stage, there is no jsp, everything is a servlet. This task will create a complete jsp page by considering all included components. Here on, the static and dynamic contents are treated differently. The resultant is a java class instead of an html page. This is how the structure of a jsp compiled into a java class will be.

2.1.2 JSP Page Compilation

The generated java servlet file is compiled into a java servlet class. The translation of a JSP source page into its implementation class can happen at any time between initial deployment of the JSP page into the JSP container and the receipt and processing of a client request for the target JSP page. The generated servlet is compiled to validate the syntax. As it is a java class, the compilation is done using javac command. This will generate the byte code to be run on JVM.

2.1.3 Class Loading

The java servlet class that was compiled from the JSP source is loaded into the container. The compiled byte code is loaded by the class loader used by web container. This is a standard process of using any java class.

2.1.4 Initialization

jspInit() method is called immediately after the instance is created. It is called only once during JSP life cycle. After the translation phase, JSP engine loads the class file and creates an instance of the servlet to handle processing of the initial request. JSP engines will call a method jspInit() to initialize the servlet. jspInit() method is generated during the translation phase which is normally used for initializing application-level parameters and resources. You can also override this method by using declaration. Initialization is done by calling the jspInit() method. This is one time activity at the start of the initialization process. Initialization will make the ServletContext and ServletConfig objects available. One can access many attributes related to the web container and the servlet itself. After initialization the servlet is ready to process requests.

```
<%!
public void jspInit()
{
//put your custom code here
}
%>
```

2.1.5 Execution Phase

After the initialization phase, the web container calls the method jspService() to handle the request and returning a response to the client. Each request handled is a separate thread. Note that, all the scriptlets and expressions end up inside this method. The JSP directives and declarations are applied to the entire page, so that they are always outside this method. In the execution phase, the container manages one or more instances of this class in response to requests and other events. The interface JspPage contains jspInit() and jspDestroy(). The JSP specification has provided a special interface, HttpJspPage for JSP pages serving HTTP requests and this interface contains jspService().

2.1.6 Service() Execution

This method is called for every request of JSP during its life cycle. Entire initialization process is done to make the servlet available in order to process the incoming request. jspService() is the method that actually processes the request. It prints the response in html (any other) format, using 'out' object. This is where it serves the purpose of creation. It has to pass through all the above steps to reach this phase. It passes the request and the response objects. jspService() cannot be overridden. jspService() is the method which is called everytime the JSP is requested to serve a request. This method normally executes in a separate thread of execution (provided we have selected Single Thread Model for the JSP) and the main JSP thread keeps waiting for other incoming requests. Everytime a request arrives, the main JSP thread spawns a new thread and passes the request (incoming request) and response (new) objects to the jspService() method which gets executed in the newly spawned thread.

2.1.7 Finalization

In the finalization phase, the web container calls the method jspDestroy(). This method is used to clean up memory and resources. Like jspInit() method, you can override the jspDestroy() method also to do your all clean up such as release the resources you loaded in the initialization phase.

```
<%!
public void jspDestroy()
{
// put your custom code here
```

```
// destroy method to clean up resources
}
%>
```

2.1.8 jspDestroy() Execution

This method is called when this JSP is destroyed. With this call, the servlet serves its purpose and submits itself to heaven (garbage collection). This is the end of JSP life cycle. jspInit(), jspService() and jspDestroy() are called the life cycle methods of the JSP. Whenever the server is shutting down or when the server needs memory, the server removes the instance of the servlet. The destroy method jspDestroy() can be called by the server after initialization and before or after request processing. Once destroyed the jsp needs to be initialized again.

Just to summarize, web container handles incoming requests to a jsp by converting it into a servlet and then by using this servlet to generate the response. Also when the server shuts down, the container needs to clear its instances.

2.2 CONTROLLING LIFE CYCLE

In above discussion, we understood that the jsp initialization phase happens when the first request hits the web container. This might take more time to present the response to the first user of jsp. It should be possible to keep initialized servlets ready to receive first request and immediately return response. This can be achieved by using the attributes provided by web and application server. The attribute can be "load-on-startup", "pre-compile", etc.

Life of the jsp page is just same as the servlet life cycle. After getting translated the jsp file is just like a servlet. The life cycle of the jsp page is given below:

jspInit() This method is called from the init() method. This method belongs to interface javax.servlet.jsp.JspPage. We can override this method. This method will be called only once when the container loads the servlet for the first time.

jspService() This method is called from the servlet's service () method. The container passes the request and response objects to this method. We cannot override this method. It is a method of javax.servlet.jsp.HttpJspPage interface.

jspDestroy() This method is called by the servlet's destroy() method. We can override this method. It is a method of javax.servlet.jsp.JspPage interface.

Example The code of the program is as follows and the output of this code is shown in Figure 2.3.

***Example* 2.1** Sample code

```
<html>
<head>
<title>The lifecycle of jsp page</title>
</head>
<body>
<h1>Showing the life cycle of jsp using jspInit and jspDestroy</h1>
<%!
int num;
int count;
public void jspInit()
{
count++;
num = 10;
}
public void jspDestroy()
{
count–;
num = 0;
}
%>
<%
```

```
out.println("The number is " + num + "<br>");
out.println("The counter is " + count + "<br>");
%>
</body>
</html>
```

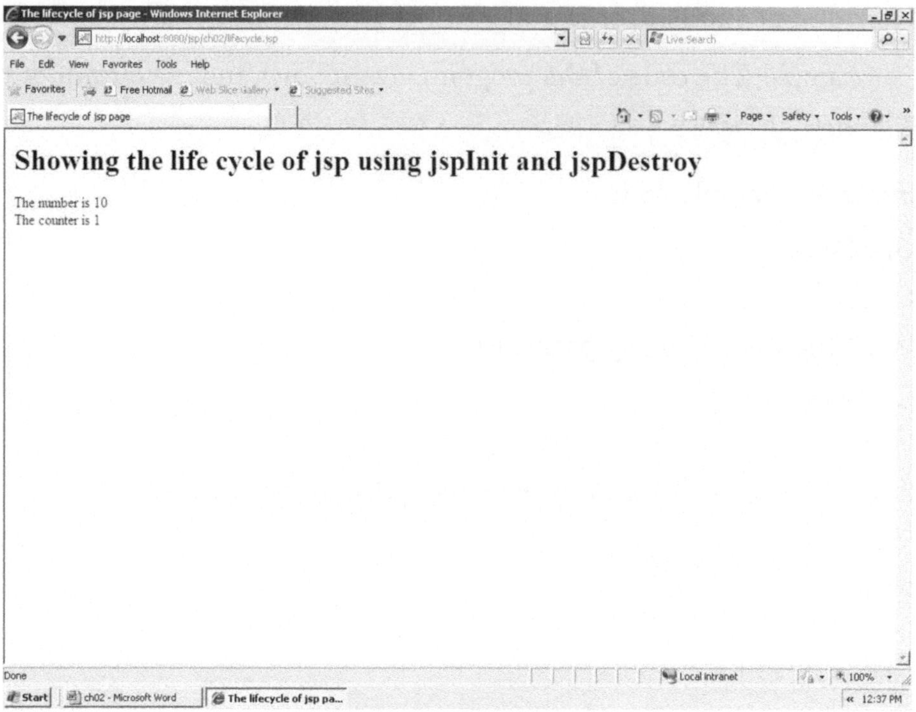

Figure 2.3 Output of sample code

2.3 LIFE CYCLE OF A SERVLET

The life cycle of a servlet can be categorized into four parts.

2.3.1 Loading and Instantiation

The servlet container loads the servlet during startup or when the first request is made. The loading of the servlet depends on the attribute <load-on-startup> of web.xml file. If the attribute <load-on-startup> has a positive value then the servlet is loaded with the container, otherwise it gets loaded when the first request comes for service. After loading of the servlet, the container creates the instances of the servlet.

2.3.2 Initialization

After creating the instances, the servlet container calls the init() method and passes the servlet initialization parameters to the init() method. The init() must be called by the servlet container before the servlet can service any request. The initialization parameters persist until the servlet is destroyed. The init() method is called only once throughout the life cycle of the servlet. The servlet will be available for service if it is loaded successfully, otherwise the servlet container unloads the servlet.

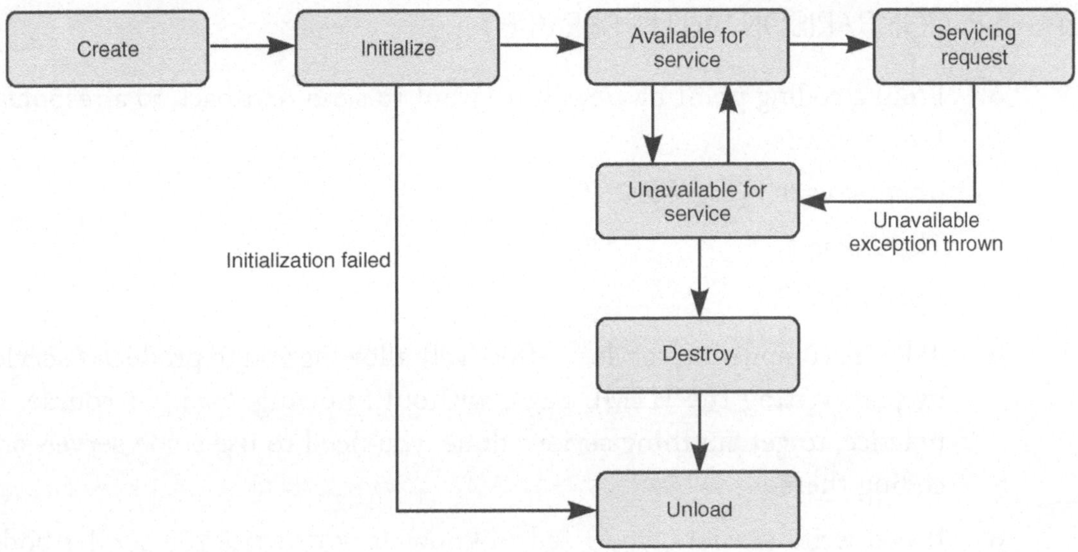

Figure 2.4 Servlet life cycle

2.3.3 Servicing the Request

After successfully completing the initialization process, the servlet will be available for service. Servlet creates seperate threads for each request. The servlet container calls the service() method for servicing any request. The service() method determines the kind of request and calls the appropriate method (doGet() or doPost()) for handling the request and sends response to the client using the methods of the response object.

2.3.4 Destroying the Servlet

If the servlet is no longer needed for servicing any request, the servlet container calls the destroy() method. Like init() method, this method is also called only once throughout the life cycle of the servlet. Calling the destroy() method indicates the servlet container not to send any request for service and the servlet releases all the resources associated with it. Java Virtual Machine (JVM) claims for the memory associated with the resources for garbage collection.

Life cycle of servlet in shown diagramatically in Figure 2.4.

2.4 JSP COMPARISON WITH SERVLET

- From a coding point of view, if we want to send text back to a response, the servlet is

  response.write("<p>Hello</P>");

  The JSP is

  <p>Hello</P>

- JSPs are compiled to servlets, effectively allowing you to produce a servlet by just writing the HTML page, without knowing Java. Of course, in practice, to get anything serious done, you need to use some server-side coding there.
- If you write servlets, you need to know Java to write the servlet body, and HTML to write the output.
- The real advantage of JSP is the use of special tags that allow you to call Java Beans directly, and the ability to produce your own custom tags (in Java) to do stuff that you cannot do with a single call to a Java Bean (e.g., inserting a list of items into the HTML, which needs a loop construct).
- So JSP allows you to put real Java directly into the HTML inside special tags.
- All this Java code will be executed before the HTML is sent to the client, and is generally used to modify the HTML page. Ironically, individuals producing JSP applications on their own often end up with a JSP containing HTML, JavaScript, and Java all in one file, which can be a little confusing, especially with all the JSP tags scattered around.

- JSP is a server-side technology that makes content generation simpler. The advantage of JSP is that they are document-centric. Servlets, on the other hand, look and act like programs. A Java Server Page can contain Java program fragments that instantiate and execute Java classes, but these occur inside an HTML template file and are primarily used to generate dynamic content. Some of the JSP functionality can be achieved on the client, using JavaScript. The power of JSP is that it is server-based and provides a framework for web application development.
- JSP does not provide any capabilities that could not in principle be accomplished with a servlet. In fact, JSP documents are automatically translated into servlets behind the scenes. But it is more convenient to write (and to modify) regular HTML than to have a zillion println statements that generate the HTML. Moreover, by separating the presentation from the content, you can put different people on different tasks—your web page design experts can build the HTML.

Servlet and JSP Sample

***Example* 2.2** Sample code of a Servlet

```
import java.io.*;
import javax.servlet.*;
import javax.servlet.http.*;
public class Hello extends HttpServlet
{
public void doGet(HttpServletRequest request, HttpServletResponse response)throws
    IOException, ServletException
  {
    response.setContentType("text/html");
    PrintWriter out = response.getWriter();
    out.println("<html>");
    out.println("<head>");
    out.println("<title>Hello Ankur!</title>");
    out.println("</head>");
```

```
            out.println("<body>");
            out.println("<h1>Hello World!</h1>");
            out.println("</body>");
            out.println("</html>");
        }
    }
```

***Example* 2.3** Sample code of JSP

```
<html>
<head>
<title>Hello Ankur</title>
</head>
<body>
<%
out.println("Hello Ankur");
%>
</body>
</html>
```

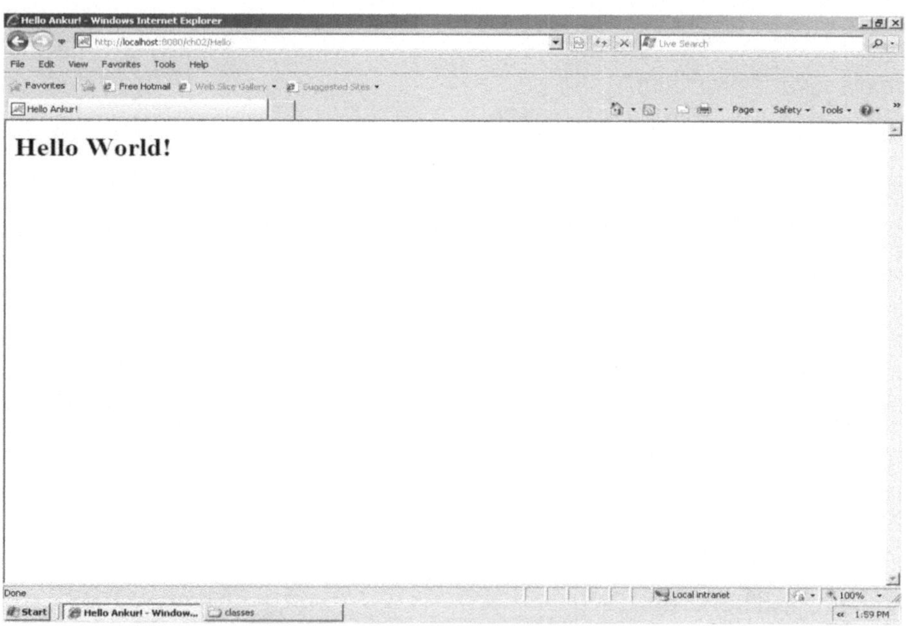

Figure 2.5 Output of Sample code of a servlet

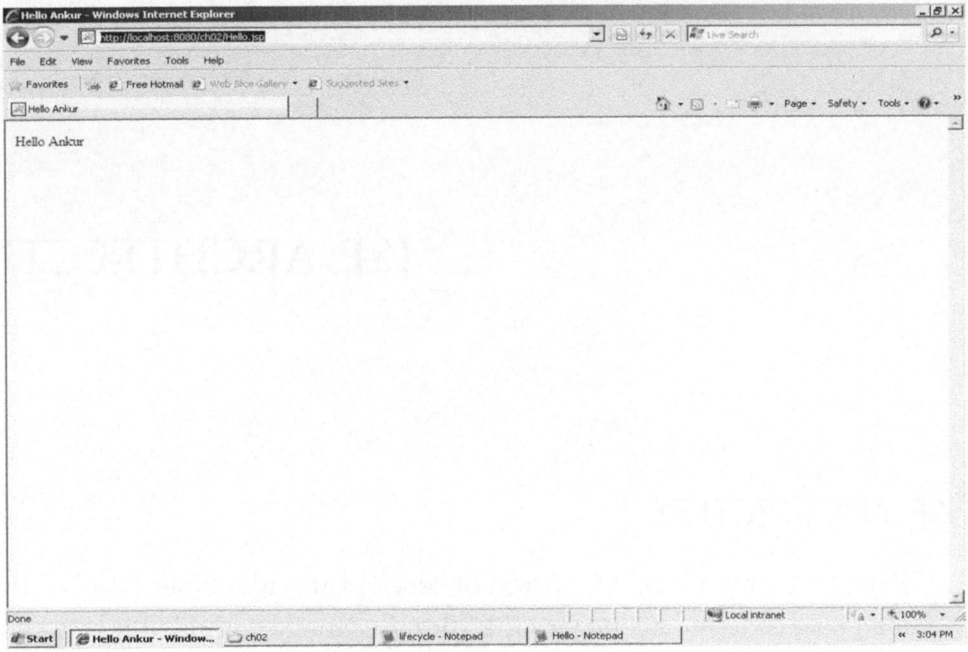

Figure 2.6 Output of Sample code of JSP

REVIEW QUESTIONS

1. How to control jsp life cycle
2. How to compare jsp with servlet.

3

JSP ARCHITECTURE

3.1 JSP ARCHITECTURE

JSP pages are high-level extension of servlets and it enables the developers to embed java code in html pages. JSP files are finally compiled into a servlet by the JSP engine, further compiled servlet is used by the engine to serve the requests. JSP is a high-end technology that helps developers insert java code in HTML pages by making use of special JSP tags. The JSPs are HTML pages but do not automatically have .html as file extension. JSP files have .jsp as extension. The following steps take place in execution of a JSP file.

i. JSP files are compiled by JSP engine into a servlet. This step creates the .jsp file as a Java servlet source file.

ii. Once this is processed, the source file above is compiled into a class file.

iii. The engine then makes use of the compiled servlet from the above process and executes requests.

Out of the three processes, the first two take time to produce a compiled servlet. This is performed only once unless modification in the source file is required. Once the compiled servlet is completed, the execution of requests is performed at a faster speed.

javax.servlet.jsp package defines two interfaces:

1. JSPPage
2. HttpJspPage

These interfaces define three methods for the compiled JSP page. The methods are:

1. jspInit()
2. jspDestroy()
3. jspService(HttpServletRequest request, HttpServletResponse response)

In the compiled JSP file these methods are present. Programmer can define only two methods, i.e., jspInit() and jspDestroy() methods.

In contrast, jspService(HttpServletRequest request,HttpServletResponse response) method is generated by the JSP engine itself (Figure 3.1).

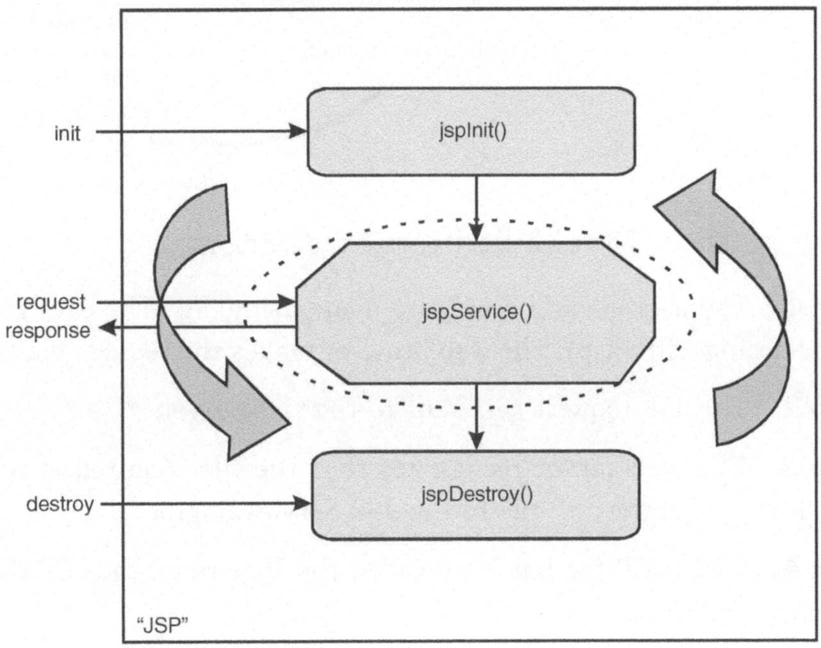

Figure 3.1 JSP architecture

3.2 STEPS REQUIRED FOR A JSP REQUEST

Figure 3.2 describes various steps involved during complete processing of a JSP request.

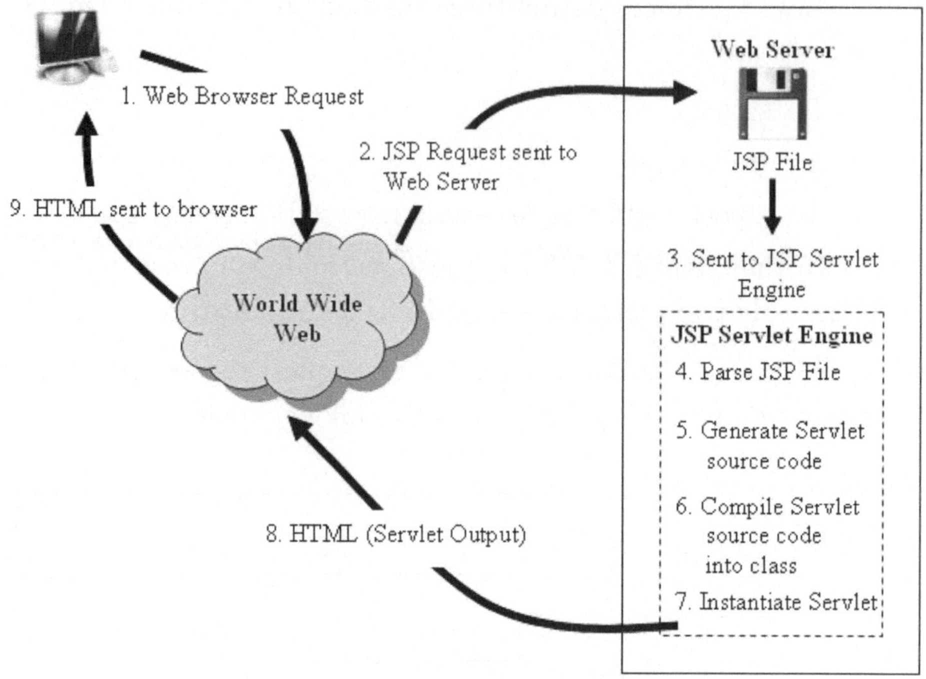

Figure 3.2 Steps required for a JSP request

Step 1 The user goes to a website made using JSP. The user goes to a JSP page (ending with .jsp). The web browser makes the request via the Internet.

Step 2 The JSP request gets sent to the web server.

Step 3 The web server recognizes that the file required is special (.jsp), therefore passes the JSP file to the JSP Servlet Engine.

Step 4 If the JSP file has been called the first time, the JSP file is parsed, otherwise go to next step.

Step 5 The next step is to generate a special Servlet from the JSP file. The entire HTML required is converted to println statements.

Step 6 The Servlet source code is compiled into a class.

Step 7 The Servlet is instantiated, calling the init and service methods.

Step 8 HTML from the Servlet output is sent via the Internet.

Step 9 HTML results are displayed on the user's web browser.

3.3 THE "PAGE-CENTRIC" APPROACH

The Page-Centric approach is also called Client-Server approach. The basic idea of Client-Server approach is that, the application lies on the client side and services the requests connecting to the server application. The steps involved in Page-Centric approach is given below.

Step 1 Client makes a request.

Step 2 The JSP page takes the request from client and processes it.

Step 3 The JSP have access to the database by using Java Beans.

Step 4 The requests are processed and the serviced results are sent back to client by JSP.

The flow of Page-Centric approach is as shown in Figure 3.3.

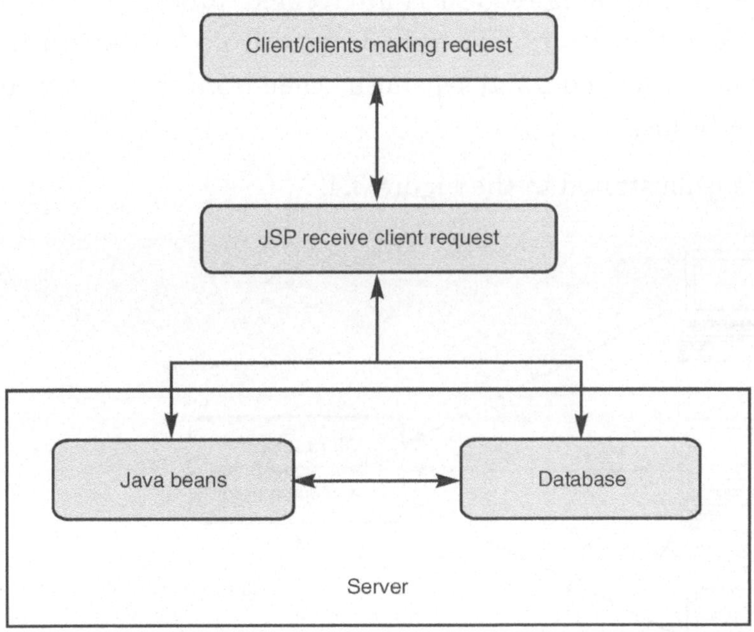

Figure 3.3 Flow of Page-Centric approach

The above approach has the advantage of simplifying the process, but its disadvantage is when the number of clients increases, the process becomes difficult to handle.

We start by looking at architectures based on the Page-Centric or Client-Server approach. We will look at the page-view and page-view with Bean architectures.

Applications built using a client-server approach have been around for some time; they consist of one or more application programs running on client machines, and connecting to a server-based application to work (A good example would be a PowerBuilder or Oracle Forms-based system.) CGIs and pre-Servlet applications were generally based on this simple 2-tier model, and with the introduction of Servlets, 2-tier applications could also be created in Java.

This model allows JSP's or Servlets direct access to some resource like a database or legacy application to service a client's request: the early JSP specifications termed this as "Model 1" programming approach. The JSP page is where the incoming request is intercepted, processed, and the response is sent back to the client; JSPs only differed from Servlets in this scenario by providing cleaner code, and separating code from the content by placing data access in Beans.

This is illustrated in the Figure 3.4.

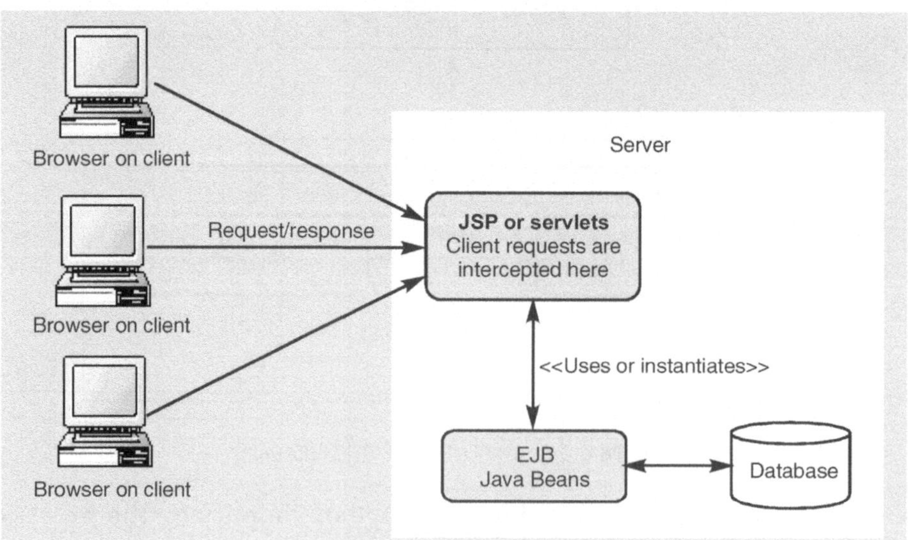

Figure 3.4 Model I approach

The advantage of such an approach is that it is simple to program, and allows the page author to generate dynamic content easily, based upon the request and the state of the resources.

However, this architecture does not scale up well for a large number of simultaneous clients since there would be a significant amount of request processing to be performed, and each request must establish or share a potentially scarce/expensive connection to the resource in question (A good example would be JDBC connections in servlets or JSPs and the need for connection pools.).

3.3.1 Page-View

This basic architecture involves direct request invocations to a server page with embedded Java code, and markup tags which dynamically generate output for substitution within the HTML.

This approach has many benefits. It is very easy to get started and is a low-overhead approach from a development standpoint. All of the Java code may be embedded within the HTML, so changes are confined to a limited area, reducing complexity. Figure 3.5 shows the architecture visually.

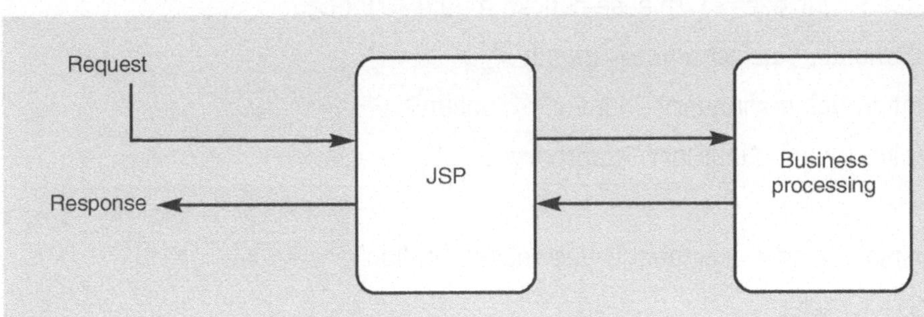

Figure 3.5 Page-view architecture

Let us begin with a basic description of the example application

Example 3.1 Pageview.jsp

```
<html>
<head>
```

```
<title>page-view</title>
</head>
<body bgcolor="#ffffff">
<form method="post" action="game.jsp" name="">
<center>
<h3>
baby game</h3> </center>
<center>
<br>
<table border cols=5 width="75%" >
<caption>please enter your name:
<input type="text" name="guesser"> </caption>
 <tr>
 <td>
<br> <input type="radio" name="gender" value="female" checked>female
<p> <input type="radio" name="gender" value="male">male
<p> </td>
 <td> <font size=-1>please choose a date:</font>
<p>month: <select name="month">
<option value="january">january</option>
<option value="february">february</option>
...
<option value="december">december</option>
<br> </select>
<p>day: <select name="day">
<option value="1">1</option>
<option value="2">2</option>
...
<option value="31">31</option>
 <br> </select>
<p> </td>
```

```
 <td> <font size=-1>please choose a weight:</font>
<p>pounds: <select name="pounds">
<option value="5">5</option>
<option value="6">6</option>
...
<option value="12">12</option>
<br> </select>
<p>ounces: <select name="ounces">
<option value="1">1</option>
<option value="2">2</option>
...
<option value="15">15</option>
 <br> </select>
<p> </td>
 <td> <font size=-1>please choose a length:</font>
<p>inches: <select name="length">
 <option value="14">14</option>
<option value="15">15</option>
...
<option value="25">25</option>
 </select> <p> <p> <p> </td>
 </tr>
</table>
</center>
 <br> 
<center>
<p> <input type="submit" name="submit" value="make guess">
          
<input type="reset" name="reset" value="reset">
</center>
 <br> </form>
```

```
</body>
</html>
```

This HTML form will POST the choices to our JSP, which is shown in source form below.

The output is shown in Figure 3.6.

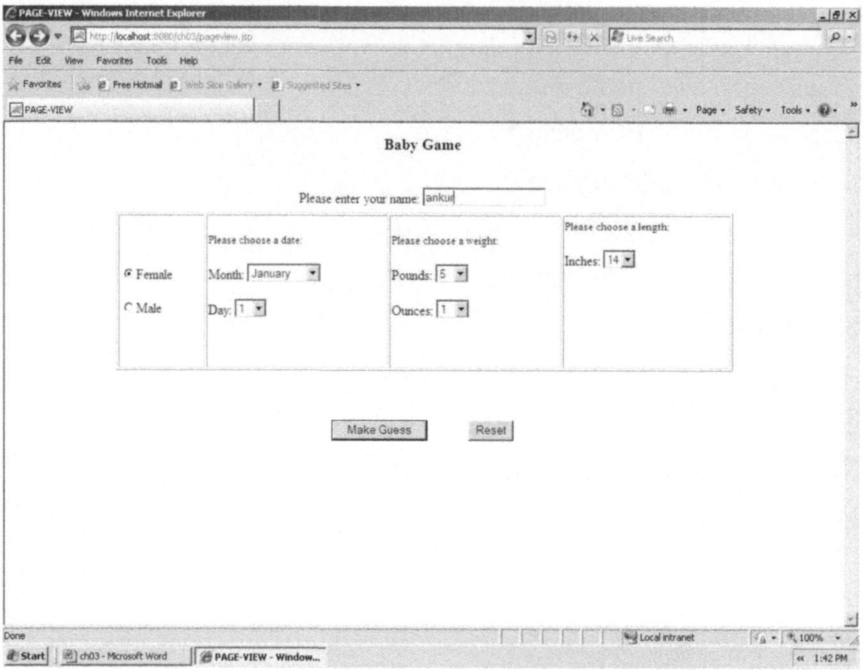

Figure 3.6 Output of pageview.jsp

game.jsp

```
<html>
<head> <title>baby game - your guesses</title></head>
<body bgcolor=ffffff>
<%@ page language="java" import="java.util.*,java.io.*" %>
<%
string guesser = request.getparameter("guesser");
string gender = request.getparameter("gender");
string pounds = request.getparameter("pounds");
```

```
string ounces = request.getparameter("ounces");
string month = request.getparameter("month");
string day = request.getparameter("day");
string length = request.getparameter("length");
if(guesser = = null || gender = = null || pounds = = null || ounces = = null
    || month = = null || day = = null || length = = null)
{ %>
<br> there where some choices that were not selected.<br> <br>
sorry, but you must complete all selections to play.<br>
<font size=-1>(please hit the browser 'back' button to continue)</font> <br>
<%}
 else
{
//store guess info and display to user
properties p = new properties();
p.put("guesser", guesser);
p.put("gender", gender);
p.put("pounds", pounds);
p.put("ounces", ounces);
p.put("month", month);
p.put("day", day);
p.put("length", length);
fileoutputstream outer = new fileoutputstream(guesser);
p.save(outer, "baby game – "+guesser+"'s guesses");
outer.flush();
outer.close();
%>
<br>
<%= guesser %>, your choices have been stored.<br>  here they are:<br>
<table border cols=5 width="75%" >
<caption> </caption>
```

```
<tr>
<td> <%= gender %>
</td>
<td>
<%= pounds %> lbs <%= ounces %> oz
</td>
<td>
<%= month %> <%= day %>
</td>
<td>
<%= length %> inches
</td>
</tr>
</table>
<br>
<%}
%>
</body>
</html>
```

This JSP certainly was easy to build and is well suited to handling such simple needs. It is getting quite cluttered with Java code, though, and its processing requirements are extremely simplistic. More sophisticated business rules and data access code would require much more Java within the JSP and the source would become unwieldy quickly as it grows. Additionally, we are limited to mostly "cut-and-paste" reuse, which means duplicate code and a more error-prone environment that is harder to maintain and debug.

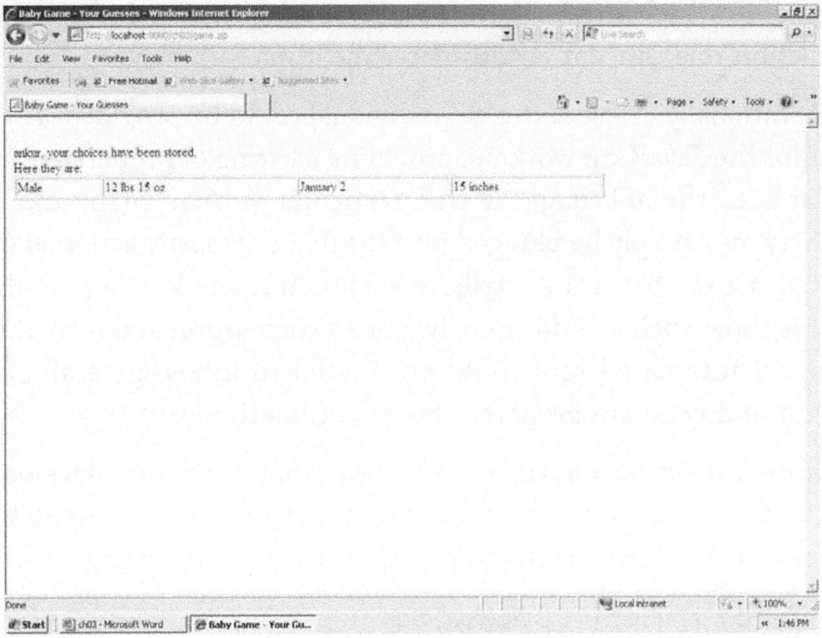

Figure 3.7 Output of Game.jsp

3.3.2 Page-View with Bean

This pattern is used when the Page-View architecture becomes too cluttered with business-related code and data access code. The Baby Game example now evolves into a more sophisticated design, as shown in the Figure 3.8.

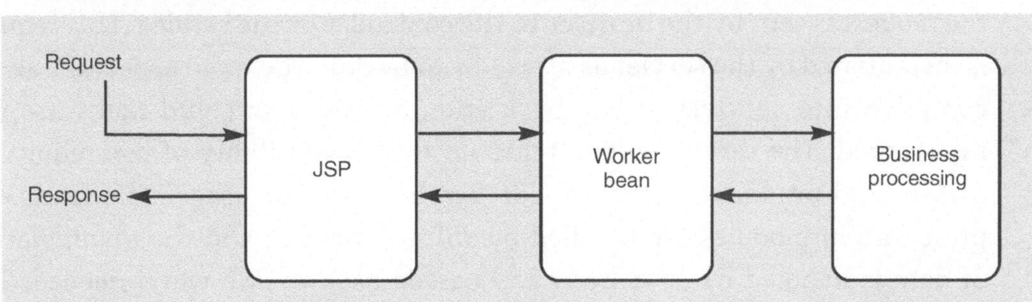

Figure 3.8 Page-view business architecture

We see in these two listings that the Java code representing the business logic and simple data storage implementation has migrated from the JSP to the JavaBean worker. This refactoring leaves a much cleaner JSP with limited

Java code, which can be comfortably owned by an individual in a web-production role, since it encapsulates most markup tags.

Additionally, a less technical individual could be provided with a property sheet for the JavaBean workers, providing a listing of properties that are made available to the JSP page by the particular worker bean, and the desired property may simply be plugged into the JSP getProperty action via the markup tag in order to obtain the attribute value. An example of a potential property sheet is shown below. Additionally, once a convention and format were agreed upon, an automated tool could be created to interrogate all JavaBeans of interest and generate property sheets automatically.

Moreover, we now have created a bean that a software developer can own, such that functionality may be refined and modified without the need for changes to the HTML or markup within the JSP source page.

3.4 THE "DISPATCHER" APPROACH

This is also called N-tier approach, where the server side of the above architecture is divided into multiple tiers, using JSP as a controller, passing requests to Java Beans.

JSP is popular because of its processing ability. Processing is distinctly divided between presentation and front components. The popular JSP Architecture is the Model View Controller (MVC) model. In this MVC model, the request is sent by the browser to the controller or the servlet. This request is instantiated by the servlet as a Java Bean by JSP. The main aspect is JSP are compiled into servlets at the back end, and the front end tasks are not interrupted. The servlet engine takes up the responsibility of compiling JSP Servlet and producing the final JSP servlet class for usage. The front end presentation modules are handled by JSP for viewing and the manipulation of data is handled by Java Bean and passed back to JSP when needed. The Presentation part of the MVC Model has no processing logic. It performs the task of extracting beans or objects that may have been initially created by the controller. It also extracts the dynamic content within it, for insertion within its static templates. The Application Manager in the MVC Model is the Controller that processes HTTP requests. They are not responsible for presentation tasks that can be either servlets or JSP. They take the task of managing the application state, security, and presentation uniformity, and thus have a single point of entry.

This explains the approach and the process of execution of a request.

We now move on to look at architectures based on the dispatcher or "N-tiered" approach, where a Servlet (or JSP) acts as a mediator or controller, delegating requests to JSP pages and JavaBeans. We will look at the mediator-view, mediator-composite view, and service to workers architectures.

In an N-tier application, the server side of the architecture is broken up into multiple tiers, as illustrated in the Figure 3.9.

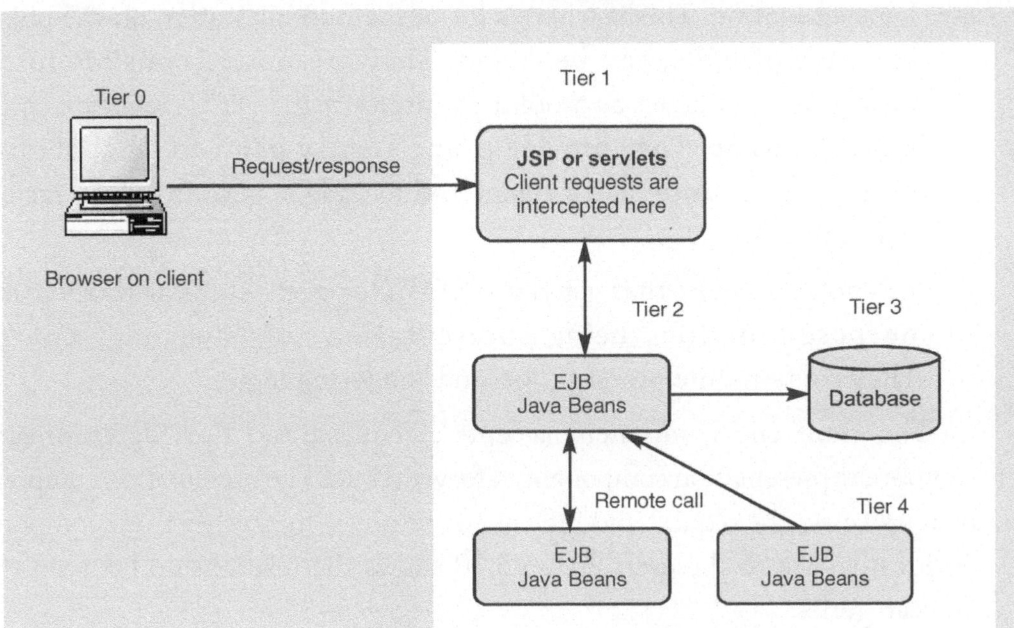

Figure 3.9 N-tier application

In this case, the application is composed of multiple tiers, where the middle tier, the JSP, interacts with the back end resources via another object or Enterprise JavaBeans component. The Enterprise JavaBeans server and the EJB provide managed access to resources, support transactions and access to underlying security mechanisms, thus addressing the resource sharing and performance issues of the 2-tier approach. This is the programming model supported by the Java 2 Enterprise Edition (J2EE) platform.

The first step in N-tiered application design should be identifying the correct objects and their interactions, in other words, object modelling. This is the part where class diagrams and tools like Rational Rose (http://www.rational.com) step in for architects. Quite often a line has to be drawn as to what objects need to be modelled: being over-zealous can cause unnecessary complexity. If you get this part right, you're half way there!

The second part is identifying the JSPs or Servlets. As a rule of thumb, these should be divided into two categories, depending on the role they play, often called "web components" in the J2EE terminology.

- Front end JSPs or Servlets that manage application flow and business logic evaluation. This is where a lot of method invocation on the objects, or usage of EJBs, can be coded. They are not responsible for any presentation, and act as a point to intercept the HTTP requests coming from the users. They provide a single entry point to an application, simplifying security management and making application state easier to maintain.
- Presentation JSPs that generate HTML (or even XML), with their main purpose in life being presentation of dynamic content. They contain only presentation and rendering logic.

The front-end component accepts a request, and then determines the appropriate presentation component to forward it to. The presentation component then processes the request and returns the response to another front component or directly back to the user: Figure 3.10 shows the relationship between these two categories.

These categories are analogous to the Model-View design pattern, where the front-end component is the model and the presentation component is the view. This approach was referred to as the "Model 2" programming model in earlier releases of the JSP specifications. It differs from Model 1 essentially in the location at which the bulk of the request processing is performed. The Figure 3.11 shows the Model 2 approach:

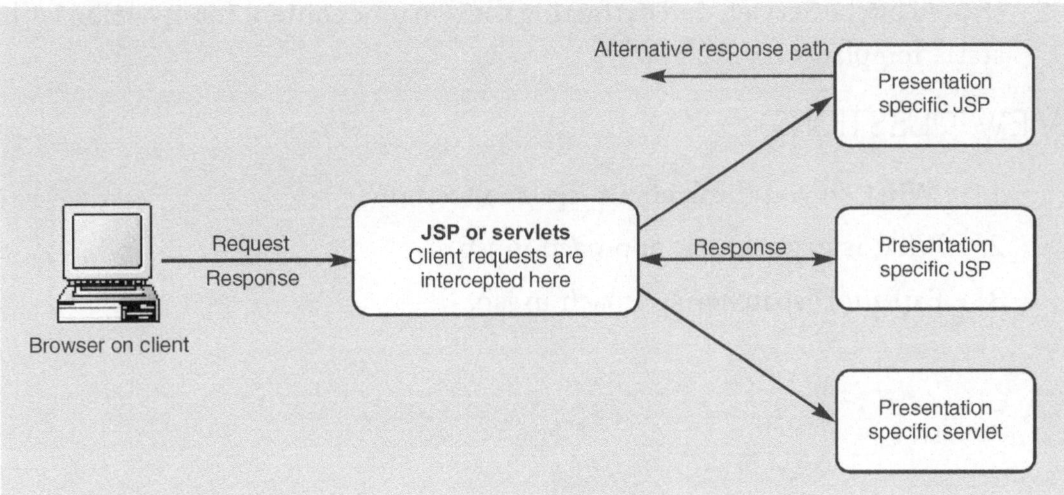

Figure 3.10 Relationship between front-end component and presentation component

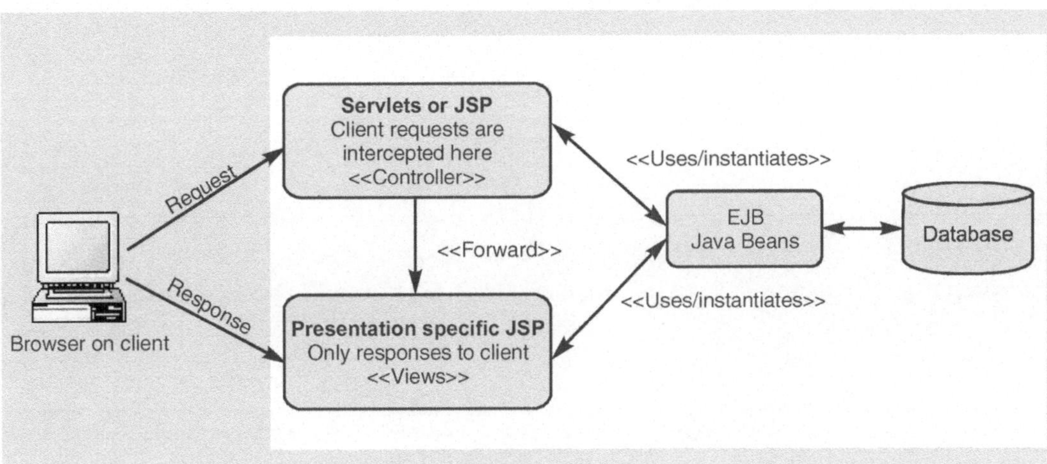

Figure 3.11 Model 2 approach

In this approach, JSPs are used to generate the presentation layer, and either JSPs or Servlets to perform process-intensive tasks. The front-end component acts as the controller and is in charge of the request processing and the creation of any beans or objects used by the presentation JSP, as well as deciding, depending on the user's actions, which JSP to forward the request. There is no processing logic within the presentation JSP itself. It is simply responsible for retrieving any objects or beans that may have been previously created by the Servlet, and extracting the dynamic content for insertion within static templates.

REVIEW QUESTIONS

1. What do you think about jsp architecture?
2. What is page centric approach in jsp.
3. Explain Dispatcher approach in jsp.

4

IMPLEMENTATION OF JSP WITH TOMCAT

4.1 INTRODUCTION TO SERVERS

A web server is a computer program that is responsible for accepting HTTP requests from clients, which are known as web browsers, and serving them HTTP responses along with optional data contents, which usually are web pages such as HTML documents and linked objects. Although web server programs differ in detail, they all share some basic common features.

4.1.1 The Web Server

A web server serves pages for viewing in a web browser, while an application server provides methods that client applications can call. A little more precisely, you can say that, a web server exclusively handles HTTP requests, whereas an application server serves business logic to application programs through any number of protocols.

A web server handles the HTTP protocol. When the web server receives an HTTP request, it responds with an HTTP response, such as sending back an HTML page. To process a request, a web server may respond with a static HTML page or image, send a redirect, or delegate the dynamic response generation to some other program such as CGI scripts, JSPs (Java Server Pages), servlets, ASPs (Active Server Pages), server-side Javascripts, or some other server-side technology. Whatever be their purpose, such server-side programs generate a response, most often in HTML, for viewing in a web browser.

Understand that a web server's delegation model is fairly simple. When a request comes into the web server, the web server simply passes the request to the program best able to handle it. The web server does not provide any functionality beyond simply providing an environment in which the server-side program can execute and pass back the generated responses. The server-side program usually provides for itself such functions as transaction processing, database connectivity, and messaging.

While a web server may not itself support transactions or database connection pooling, it may employ various strategies for fault tolerance and scalability such as load balancing, caching, and clustering features oftentimes erroneously assigned as features reserved only for application servers.

4.1.2 The Application Server

As for the application server, according to our definition, an application server exposes business logic to client applications through various protocols, possibly including HTTP. While a web server mainly deals with sending HTML for display in a Web browser, an application server provides access to business logic for use by client application programs. The application program can use this logic just as it would call a method on an object.

Such application server clients can include GUIs (graphical user interface) running on a PC, a web server, or even other application servers. The information travelling back and forth between an application server and its client is not restricted to simple display markup. Instead, the information is program logic. Since the logic takes the form of data and method calls, and not static HTML, the client can employ the exposed business logic however it wants.

In most cases, the server exposes this business logic through a component API, such as the EJB (Enterprise JavaBean) component model found on J2EE (Java 2 Platform, Enterprise Edition) application servers. Moreover, the application server manages its own resources. Such gate-keeping duties include security, transaction processing, resource pooling, and messaging. Like a web server, an application server may also employ various scalability and fault-tolerance techniques.

4.2 INTRODUCTION OF TOMCAT

Tomcat is an application server from the Apache Software Foundation that executes Java servlets and renders web pages that include Java Server Page coding. It is described as a "reference implementation" of the Java Servlet and the Java Server Page specifications, Tomcat is the result of an open collaboration of developers and is available from the Apache website in both binary and source versions. Tomcat can be used as either a standalone product with its own internal web server or together with other web servers, including Apache, Netscape Enterprise Server, Microsoft Internet Information Server (IIS) and Microsoft Personal Web Server. Tomcat requires a Java Runtime Enterprise Environment that conforms to JRE 1.1 or later. Tomcat is one of several open-source collaborations that are collectively known as Jakarta. It adds tools for configuration and management, but can also be configured by editing configuration files that are normally XML-formatted. Tomcat includes its own internal HTTP server.

The following options are available for tomcat.

Introduction A brief, high level, overview of Apache Tomcat.

Setup How to install and run Apache Tomcat on a variety of platforms.

First web application An introduction to the concepts of a web application as defined in the Servlet 2.4 specification, covers basic organization of your web application source tree, the structure of a web application archive, and an introduction to the web-application deployment descriptor (/WEB-INF/web.xml).

Deployer Operating the Apache Tomcat Deployer to deploy, precompile, and validate web applications.

Manager Operating the Manager web app to deploy, undeploy, and re-deploy applications while Apache Tomcat is running.

Realms and access control Description of how to configure Realms (databases of users, passwords, and their associated roles) for use in web applications that utilize Container Managed Security.

Security manager Configuring and using a Java Security Manager to support fine-grained control over the behaviour of your web applications.

JNDI resources Configuring standard and custom resources in the Java Naming Director Interface (JNDI) naming context that is provided to each web application.

JDBC DataSource Configuring a JNDI DataSoure with a DB connection pool, examples for many popular databases.

Class loading Information about class loading in Apache Tomcat 6, including where to place your application classes so that they are visible.

JSPs Information about Jasper configuration, as well as the JSP compiler usage.

SSL Installing and configuring SSL support so that your Apache Tomcat will serve requests using the https protocol.

SSI Using Server Side Includes in Apache Tomcat.

CGI Using Common Gateway Interfaces with Apache Tomcat.

Proxy support Configuring Apache Tomcat 6 to run behind a proxy server (or a web server functioning as a proxy server).

MBean descriptor Configuring MBean descriptor files for custom components.

Default servlet Configuring the default servlet and customizing directory listings.

Apache Tomcat clustering Enable session replication in a Apache Tomcat environment.

Balancer Configuring, using, and extending the load balancer application.

Connectors Connectors available in Apache Tomcat, and native web server integration.

Monitoring and management Enabling JMX Remote support, and using tools to monitor and manage Apache Tomcat.

Logging Configuring logging in Apache Tomcat.

Apache Portable Runtime Using APR to provide superior performance, scalability and better integration with native server technologies.

Virtual hosting Configuring vitual hosting in Apache Tomcat.

Advanced IO Extensions available over regular, blocking IO (Input/Output).

Additional components Obtaining additional, optional components.

4.2.1 Running Tomcat as a Windows Service

In a production environment, it is more convenient to run Tomcat as a "service", so that it can start automatically whenever the system is started (or re-start automatically after an unexpected interruption).

To install Tomcat as a service, start a CMD shell (with administrator right) and run

Change directory to $CATALINA_HOME\BIN

$CATALINA_HOME\BIN> SERVICE INSTALL

Installing the service 'Tomcat6'

Using CATALINA_HOME: D:\TOMCAT-6.0.20

Using CATALINA_BASE: D:\TOMCAT-6.0.

Using JAVA_HOME: D:\JDK1.6

Using JVM: D:\JDK1.6\JRE\BIN\SERVER\JVM.DLL

The Tomcat service called "Apache Tomcat 6" is installed and will start automatically whenever the system is started. Check the "Services" under "Control Panel" , "Administrative Tools".

4.2.2 Tomcat's Startup Script

To start Tomcat server, you could invoke the batch file "startup.bat" (in directory "$CATALINA_HOME\bin", where CATALINA_HOME refers to the Tomcat installed directory). The "startup.bat" invokes "catalina.bat start".

Alternatively, you could call the "catalina.bat" directly, which provides more options of starting Tomcat. Enter "catalina" to view the options.

catalina

Using CATALINA_BASE:	D:\BIN\TOMCAT6.0
Using CATALINA_HOME:	D:\BIN\TOMCAT6.0
Using CATALINA_TMPDIR:	D:\BIN\TOMCAT6.0.16\TEMP
Using JRE_HOME:	D:\BIN\JDK1.6
Usage:	catalina (commands)

Command

debug	Start Catalina in a debugger
debug	Security Debug Catalina with a security manager
jpda start	Start Catalina under Java Platform Debugger Architecture (JPDA) debugger
run	Start Catalina in the current window
run-security	Start in the current window with security manager
start	Start Catalina in a separate window
start-security	Start in a separate window with security manager
stop	Stop Catalina
version	What version of tomcat are you running?

4.2.3 Tomcat Security

The Tomcat also provide container managed security for the web application deployed in the server. The Tomcat has a xml file, conf/tomcat-user.xml, which contains all the username, password and roles for the server. The file is given below. We need to add our username and password which can be used to validate before accessing any page for the application. Like for example, we have added admin role with visualbuilder username.

Example 4.1 tomcat-user.xml file

```
<?xml version = '1.0' encoding = 'utf-8'?>
<tomcat-users>
```

```
    <role rolename = "tomcat"/>
    <role rolename = "role1"/>
    <role rolename = "admin"/>
    <user username = "tomcat" password = "tomcat" roles = "tomcat"/>
    <user username = "both" password = "tomcat" roles = "tomcat,role1"/>
    <user username = "role1" password = "tomcat" roles = "role1"/>
    <user username = "visualbuilder" password = "test" roles = "admin"/>
</tomcat-users>
```

4.3 STEPS TO INSTALL TOMCAT

Following are the steps for Tomcat installation.

Step 1 When you click on the page, "http://tomcat.apache.org/download-60.cgi".

The **File Download-Security Warning** dialog box (Figure 4.1) appears.

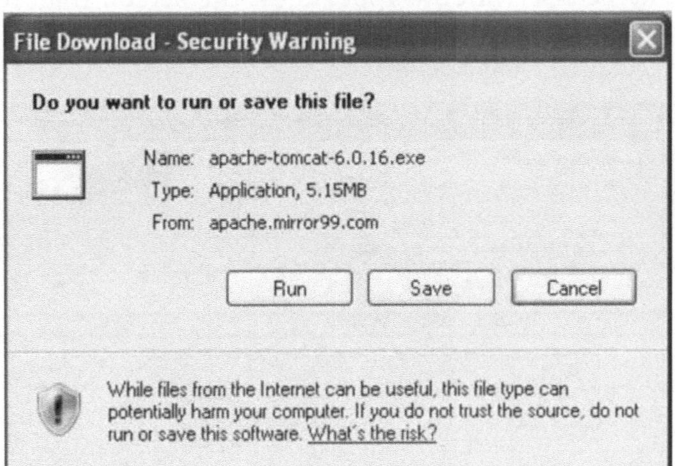

Figure 4.1 File download security warning

In windows Service Installer (pgp, md5), following screen would show up, you can either save it first and then install it or click on Run to install.

Step 2 After download, execute the file. The **Apache Tomcat Setup** dialog box (Figure 4.2) appears. Now click "Next".

Figure 4.2 Apache tomcat setup dialog box

Step 3 The license window appears on the screen (Figure 4.3). Then click "I Agree".

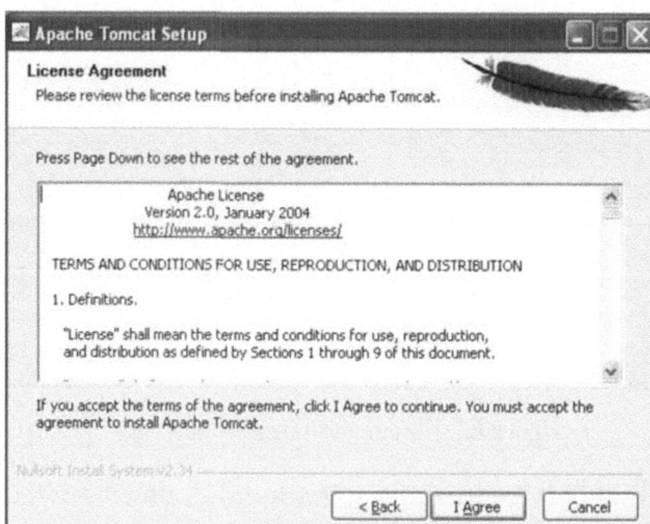

Figure 4.3 Licence agreement dialog box

Step 4 In the choose component Dialog box (Figure 4.4) select Full in the combo box and click "Next".

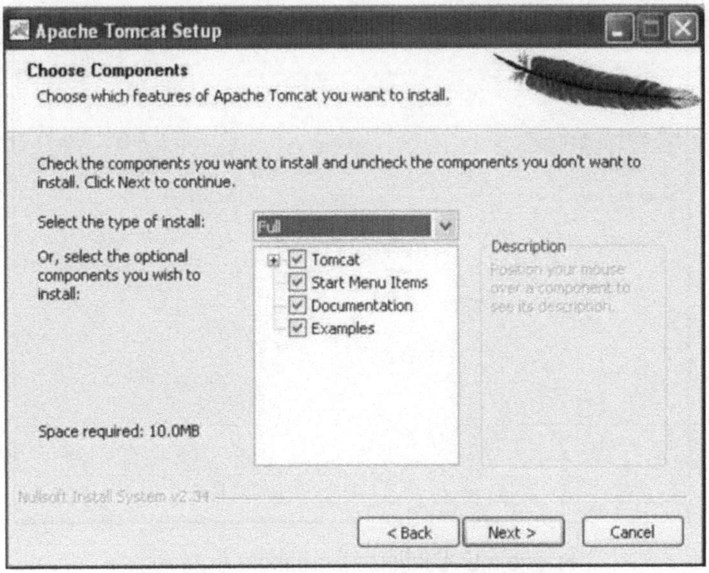

Figure 4.4 Choose components dialog box

Step 5 Choose install location and click Next

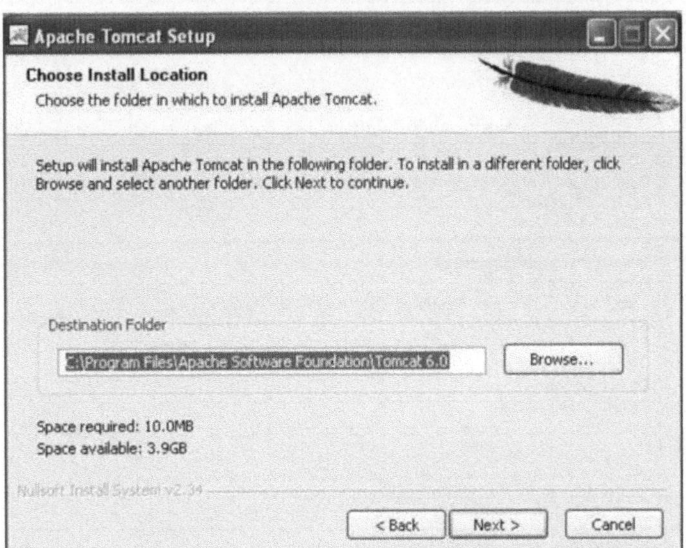

Figure 4.5 Choose install location dialog box

Step 6 Setup tomcat configuration and Click Next

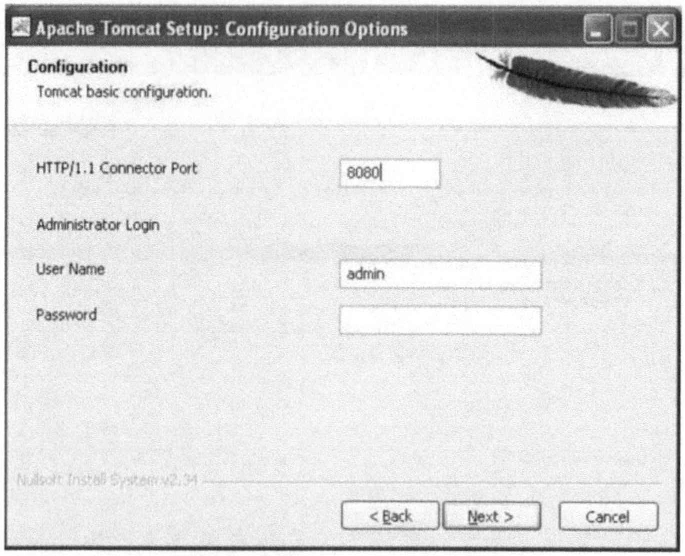

Figure 4.6 Configuration options dialog box

Step 7 The following screen Figure 4.7 would show installation is in progress.

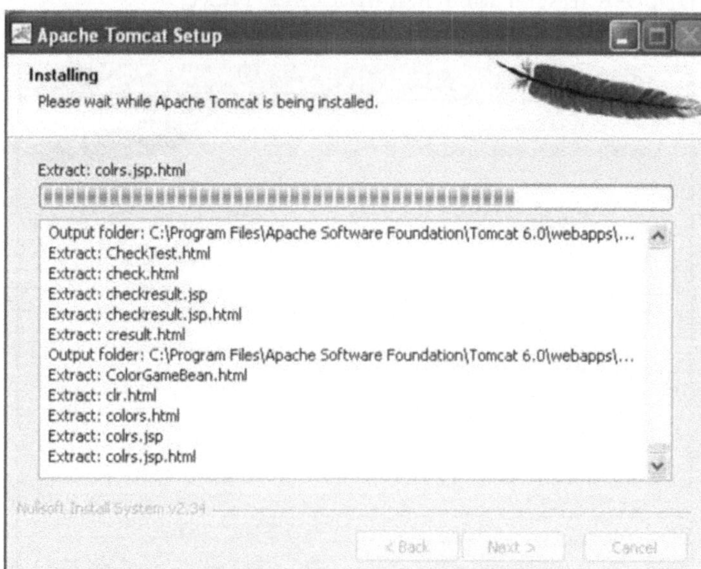

Figure 4.7 Installation in progress

Step 8 Click Finish. The complete tomcat setup Wizard should look like this.

Figure 4.8 Set up completion wizard

4.4 DIRECTORY STRUCTURE UNDER TOMCAT SERVER

After installing tomcat, you should have the following directory structure under "jakarta.tomcat". Figure 4.9 gives the directory structure of a tomcat server.

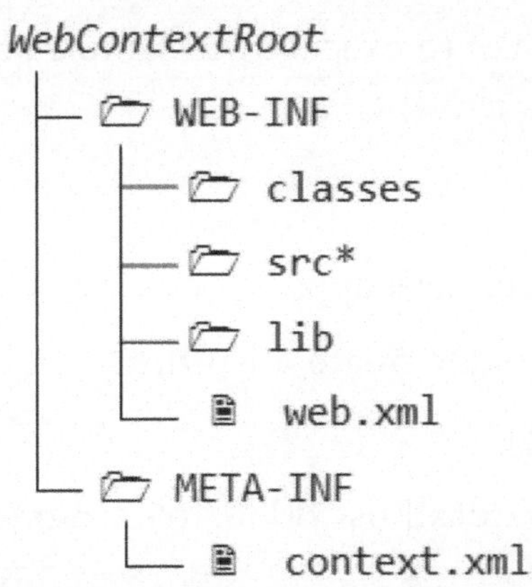

Figure 4.9 Directory structure under tomcat

Table 4.1 Tomcat directary

Directory	Purpose
/bin	This directory contains the startup and shutdown scripts for both Windows and Linux
/conf	This directory contains the main configuration files for Tomcat. The two most important are the server.xml and the global web.xml
/server	This directory contains the Tomcat Java Archive files.
/lib	This directory contains Java Archive files that Tomcat is dependent upon.
/logs	This directory contains Tomcat's log files.
/src	This directory contains the source code used by the Tomcat server. Once Tomcat is released, it will probably contain interfaces and abstract classes only.
/webapps	All web applications are deployed in this directory; it contains the WAR file.
/work	This is the directory in which Tomcat will place all servlets that are generated from JSPs. If you want to see exactly how a particular JSP is interpreted, look in this directory.

4.5 STEPS FOR EXECUTING JSP FILE

The steps you must take to execute the sample JSP into tomcat are:

- Configure tomcat6.0
- Start server
- Create folder inside webapps
- Inside webapps folder create a jsp file
- Open the browser
- Then type http://localhost:8080/foldername/jspfile

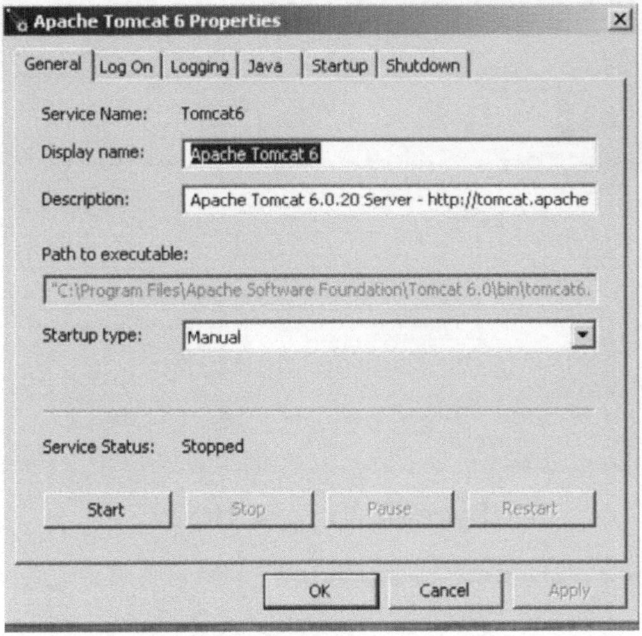

Figure 4.10 Tomcat property window

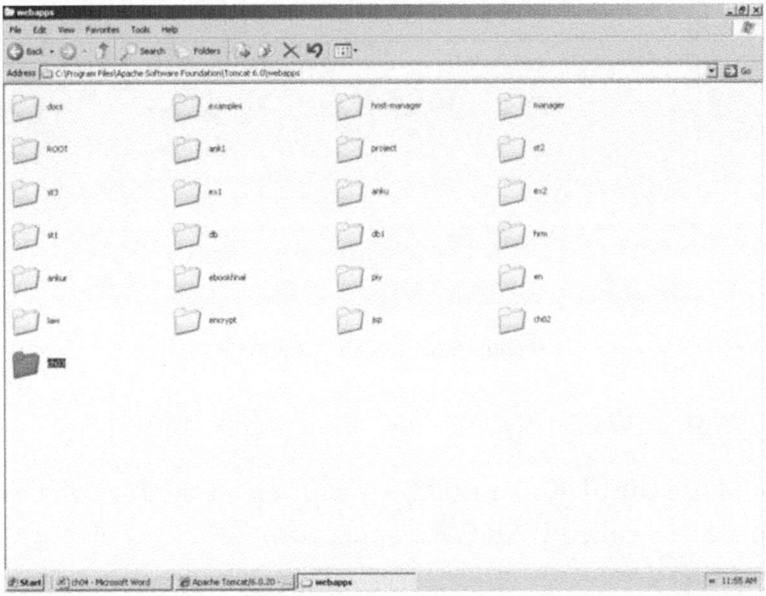

Figure 4.11 Root of tomcat

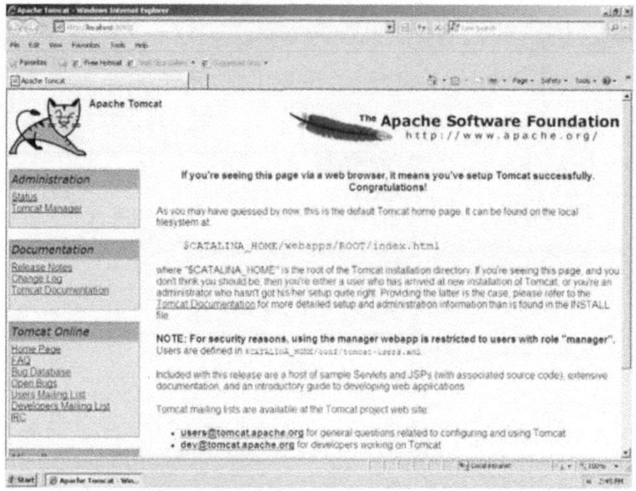

Figure 4.12 Homepage of tomcat

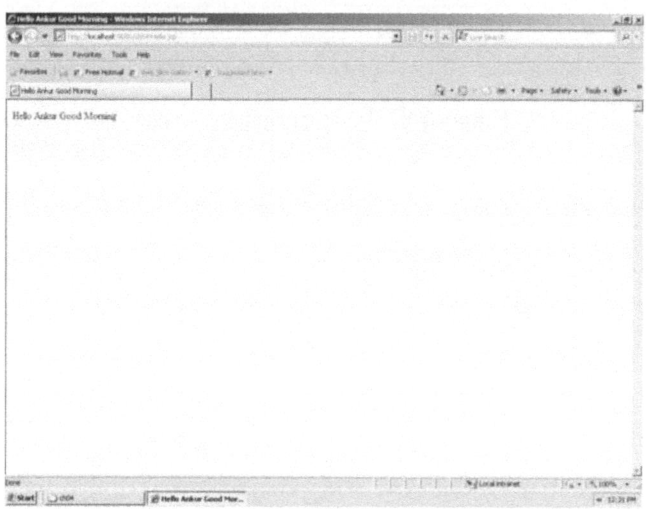

Figure 4.13 Output of Hello.jsp

4.5.1 Deploying a Web Application in a WAR file

You could use the JDK's jar utiltiy to "zip" up all the files of a web application to produce a so-called WAR (Web application ARchive) file for deployment, or distribution.

Change current directory to the web application's context root

```
> JAR CVF CH04.WAR
```

Drop the ch04.war into $CATALINA_HOME\WEBAPPS. A context called test will be created automatically. You can access the web application via URL, http://host:port/ch04.

(Tomcat actually unpacks the ch04.war into a test directory in $CATALINA_HOME\webapps.)

REVIEW QUESTIONS

1. What is the difference between web server and application server?
2. What is tomcat?
3. How to install tomcat in your pc.
4. What is the directory structure of tomcat?
5. How to run your first jsp under tomcat.
6. How to deploy a web server under tomcat.

5
STARTING WITH JSP

5.1 INTRODUCTION

JSP for short is Sun's solution for developing dynamic websites. JSP provides excellent server-side scripting support for creating database-driven web applications. JSP enable the developers to directly insert java code into jsp file, this makes the development process very simple and its maintenance also becomes very easy. JSP pages are efficient, it loads into the web server's memory on receiving the request very first time and the subsequent calls are served within a very short period of time.

5.2 JSP SCRIPTING ELEMENTS

Declarations, scriptlets and expressions are called JSP scripting elements. A JSP file with only static HTML content is also a valid JSP page, but it will have no dynamic content. It is the scripting elements which gives the JSP the capability of dynamic content generation.

Comment Documents the JSP page but is not inserted into the response.

JSP Syntax <%- - comment - -%>

Declaration Declares a variable or method valid in the scripting language used in the JSP page.

JSP Syntax <%! declaration; [declaration;]+ ... %>

Expression Contains an expression valid in the scripting language used in the JSP page.

JSP Syntax `<%= expression %>`

Scriptlet Contains a code fragment valid in the page scripting language.

JSP Syntax `<% code fragment %>`

EL Expression Contains an expression in the JSP Expression Language (EL). It allows easy access to application data stored in JavaBeans components.

a literal ${ by '${'

5.3 JSP DIRECTIVE

The directive tag gives special information about the page to JSP Engine. This changes the way JSP Engine processes the page. Using directive tag, user can import packages, define error handling pages or session information of JSP page.

Attribute directive Declares attributes of custom actions defined in tag files and are used in tag files only.

JSP Syntax

```
<%@ attribute   name="attribute-name"
  [ required="true | false" ]
  [ fragment="true | false" ]
  [ rtexprvalue="true | false" ]
  [ type="java.lang.String | a non-primitive type"]
  [ description="text" ]
%>
```

Include directive Includes a static file in a JSP page, parsing the file's JSP elements.

JSP Syntax

```
<%@ include file="relativeURL" %>
```

Page directive Defines attributes that apply to an entire JSP page.

JSP Syntax

```
<jsp:directive.page pageDirectiveAttrList />
```

Tag directive Used for declaring custom tag properties.

JSP Syntax

<jsp:directive.tag *tagDirectiveAttrList* />

where, *tagDirectiveAttrList* is the same as the attribute list in the JSP syntax.

Taglib directive Defines a tag library and prefix for the custom tags used in the JSP page.

JSP Syntax

<%@ taglib {uri= "*URI* " | tagdir="/WEB-INF/tags[/*subdir*]+"}
prefix= "*tagPrefix* " %>

Note: URI stands for Uniform Resource Indentifier.

Variable directive Declares an expression language variable exposed by the tag to the calling page.

JSP Syntax

<jsp:directive.variable variableDirectiveAttrList />

5.4 STANDARD ACTIONS

Standard action elements basically are tags that can be embedded into a jsp page. At compile time they are replaced by java code that corresponds to predefined tasks.

<jsp:attribute> The jsp:attribute element allows you to define the value of a tag attribute in the body of an XML element instead of the value of an XML attribute.

JSP Syntax

<jsp:attribute name="*attributeName*" [trim= "**true** | false"] />

<jsp:body> Specifies the body of a tag.

JSP Syntax

<jsp:body>any elements or text</jsp:body>

<jsp:element> Dynamically generates an XML element.

JSP Syntax

```
(<jsp:element name="elementName"
 ( /> | > ( any elements or text </jsp:element> ))
)
| ( <jsp:element name="elementName">
      [ <jsp:attribute name="attributeName"
        [ trim="true | false" ]
     ( /> | (any elements or text </jsp:attribute> ) ) ]+
     [ <jsp:body> any elements or text </jsp:body> ]
   </jsp:element>
 )
```

<jsp:doBody> Evaluates the body of the tag used by the calling page to invoke this tag file.

JSP Syntax

```
<jsp:doBody ({var="scopedAttributeName" |
   varReader="scopedAttributeName" }
  [scope="page | request | session | application" ] />) | />
```

<jsp:forward> Forwards a request to a web resource.

JSP Syntax

```
<jsp:forward page="{relativeURL | '${' Expression'}' |
  <%= expression %>}" { /> | > [<jsp:param name="parameterName"
value="{parameterValue | '${' Expression '}' | <%= expression %>}" }
/> ]+ </jsp:forward>}
```

<jsp:getProperty> Inserts the value of a bean property into the result.

JSP Syntax

```
<jsp:getProperty name="beanInstanceName" property="propertyName" />
```

<jsp:include> Includes a static file or the result from another web component.

JSP Syntax

```
<jsp:include page="{relativeURL | '${' Expression '}' |
```

```
  <%= expression %>}"
  [ flush="true| false" ]
  { /> | > [ <jsp:param name="parameterName"
    value="{parameterValue | '${' Expression '}' |
    <%= expression %>}" />] +
</jsp:include> }
```

<jsp:invoke> Evaluates a fragment attribute.

JSP Syntax

```
<jsp:invoke fragment="fragmentName"
({var="scopedAttributeName" |
varReader="scopedAttributeName"}
[scope="page | request | session | application" ] />) | />
```

<jsp:output> Specifies the XML declaration or the document type declaration in the request output of a JSP document or a tag file that is in XML syntax.

XML Syntax

```
<jsp:output ( omit-xml-declaration="yes|no|true|false" )
  { doctypeDecl } />
  doctypeDecl ::= ( doctype-root-element="rootElement"
  doctype-public="PubidLiteral"
  doctype-system="SystemLiteral" )
  | ( doctype-root-element="rootElement"
    doctype-system="SystemLiteral" )
```

<jsp:plugin> Causes the execution of an applet or bean. The applet or bean executes in the specified plugin. If the plugin is not available, displays a dialog to initiate the download of the plugin software.

JSP Syntax

```
<jsp:plugin
  type="bean|applet"
  code="classFileName"
```

```
codebase="classFileDirectoryName"
[ name="instanceName" ]
[ archive="URIToArchive, ..." ]
[ align="bottom|top|middle|left|right" ]
[ height="{displayPixels | <%= expression %>}"]
[ width="{displayPixels | <%= expression %>}"]
[ hspace="leftRightPixels" ]
[ vspace="topBottomPixels" ]
[ jreversion="JREVersionNumber | 1.2" ]
[ nspluginurl="URLToPlugin" ]
[ iepluginurl="URLToPlugin" ] >
   [ <jsp:params>
  [ <jsp:param name="parameterName"
    value="{parameterValue | '${' Expression '}' |
    <%= expression %>}" /> ]+
</jsp:params> ]
     [ <jsp:fallback> text message if plugin download fails
  </jsp:fallback> ]
   </jsp:plugin>
```

<jsp:root> Defines standard JSP elements and namespace attributes of tag libraries. .

<jsp:setProperty> Sets a property value or values in a bean.

JSP Syntax

```
<jsp:setProperty name="beanInstanceName"
{
  property="*" |
  property="propertyName" [ param="parameterName" ] |
  property="propertyName" value="{stringLiteral|
    '${' Expression '}' | <%= expression %>}"
}
/>
```

<jsp:text> Encloses template data.

JSP Syntax

```
<jsp:text>
  template data
</jsp:text>
```

<jsp:useBean> Locates or instantiates a bean with a specific name and scope.

JSP Syntax

```
<jsp:useBean id="beanInstanceName"
  scope="page|request|session|application"
{
  class="package.class" [ type="package.class" ]|
  beanName="{package.class | '${' Expression '}' |
    <%= expression %>}" type="package.class" |
  type="package.class"
}
{ /> | > other elements </jsp:useBean> }
```

5.5 IMPLICIT OBJECTS

Implicit objects in jsp are the objects that are created by the container automatically and the container makes them available to the developers, the developer need not create them explicitly. Since these objects are created automatically by the container and are accessed using standard variables; hence, they are called implicit objects. The implicit objects are parsed by the container and inserted into the generated servlet code. They are available only within the jspService method and not in any declaration. Implicit objects are used for different purposes. Our own methods (user-defined methods) cannot access them as they are local to the service method and are created at the conversion time of a jsp into a servlet.

request Request object has a request scope that is used to access the HTTP request data, and also provides a context to associate the request-specific data. Request object implements javax.servlet.ServletRequest interface. It uses the getParameter() method to access the request parameter. The container passes this object to the jspService() method.

response This object has a page scope that allows direct access to the HTTPServletResponse class object. "Response" object is an instance of the classes that implements the javax.servlet. ServletResponse class. Container generates this object and passes the jspService() method as a parameter.

out This object allows us to access the servlet's output stream and has a page scope. "Out" object is an instance of javax.servlet.jsp.JspWriter class. It provides the output stream that enable access to the servlet's output stream.

session "Session" object has a session scope that is an instance of javax.servlet.http.HttpSession class. Perhaps it is the most commonly used object to manage the state contexts. This object persist information across multiple user connection.

application These objects have an application scope. These objects are available at the widest context level, that allows to share the same information between the JSP page's servlet and any web components within the same application.

config These object have a page scope and is an instance of javax.servlet.ServletConfig class. "config" object allows to pass the initialization data to a JSP page's servlet. Parameters of these objects can be set in the deployment descriptor (web.xml) inside the element <jsp-file>. The method getInitParameter() is used to access the initialization parameters.

pageContext pageContext has a page scope. "pageContext" is the context for the JSP page itself that provides a single API to manage the various scope attributes instead of scoped attributes and discussed in next chapter. This API is extensively used if we are implementing JSP custom tag handlers. pageContext also provides access to several page attributes like including some static or dynamic resource.

page This object has a page scope and is an instance of the JSP page's servlet class that processes the current request. "Page" object represents the current page that is used to call the methods defined by the translated servlet class. First, typecast the servlet before accessing any method of the servlet through the page.

exceptions This object has a page scope and is an instance of java.lang.Throwable class. This object allows the exception data to be accessed only by designated JSP "error pages."

5.6 JSP SESSION

In session management whenever a request comes for any resource, a unique token is generated by the server and transmitted to the client by the response object and stored on the client machine as a cookie. We can also say that the process of managing the state of a web-based client is through the use of session IDs. Session IDs are used to uniquely identify a client browser, while the server- side processes are used to associate the session ID with a level of access. Thus, once a client has successfully authenticated to the web application, the session ID can be used as a stored authentication voucher so that the client does not have to re-type their login information for each page request. Now, whenever a request goes from this client, again the ID or token will also be passed through the request object, so that the server can understand from where the request is coming.

5.7 JSP BEAN

To use a Bean in a JSP page, three attributes must be supplied–an id, which provides a local name for the Bean, the Bean's classname, which is used to instantiate the Bean if it does not exit, and a scope, which specifies the lifetime of the Bean.

There are four scopes available–page, request, session, and application. A page-scoped bean is available only within the JSP page and is destroyed when the page has finished generating its output for the request. A request-scoped bean is destroyed when the response is sent. A session-scoped bean is destroyed when the session is destroyed. An application-scoped bean is destroyed when the web application is destroyed.

5.8 CREATE ENVIRONMENT VARIABLE

Create an environment variable called "JAVA_HOME" and set it to the JDK installed directory (e.g., c:\program files\java\jdk1.6).

5.8.1 In Windows XP

- Click Start button
- goto Control Panel
- Choose System

5.8.2 In (Vista/Windows 7)

- Advanced system settings
- Advanced tab
- Environment Variables
- System Variable
- Choose New
- Enter "JAVA_HOME" in the "Variable Name" and your JDK base directory (e.g., "C:\program file\java\jdk1.6.0") in the "Variable Value".
- To verify, start a cmd shell and issue.

REVIEW QUESTIONS

1. What is scripting element in jsp?
2. What is the concept of jsp directive?
3. What are standard actions in jsp?
4. What are implicit object in jsp.
5. What is role of environment variable?

6
JSP SCRIPTING ELEMENTS

6.1 INTRODUCTION

JSP scripting elements are used to create and access objects, define methods, and manage the flow of control. Since one of the goals of JSP technology is to separate static template data from the code needed to dynamically generate content, very sparing use of JSP scripting is recommended. Much of the work that requires the use of scripts can be eliminated by using custom tags.

JSP technology allows a container to support any scripting language that can call Java objects. If you wish to use a scripting language other than the default java, you must specify it in a page directive at the beginning of a JSP page.

```
<%@ page language="scripting language" %>
```

Since scripting elements are converted to programming language statements in the JSP page's servlet class, you must import any classes and packages used by a JSP page. If the page language is java, you import a class or package with the page directive.

```
<%@ page import="packagename.*, fully_qualified_classname" %>
```

Syntax of JSP scripting language

```
<%
//java codes
%>
```

JSP Engine places these codes in the jspService() method.

The following codes show the demonstration of scripting language.

```
<%
//java codes
String endusers=null;
userName=request.getParameter("endusers");
%>
```

6.2 COMMENT

A comment, marks text or lines that the JSP container should ignore. A comment is useful when you want to comment out part of your JSP page. The JSP container does not process anything within the <%-- and --%> characters or within the <!-- and --> characters. A comment is not inserted into the response.

JSP syntax

```
<%– comment --%>
```

XML syntax

```
<!-- comment -->
```

***Example* 6.1** *comment.jsp*

This example display comment in JSP

```
<%@ page language="java" %>
<html>
<head> <title>A Comment Test</title> </head>
<body>
<h2>A Test of Comments</h2>
<%-- This comment will not be included in the response --%>
</body>
</html>
```

The output is given in Figure 6.1.

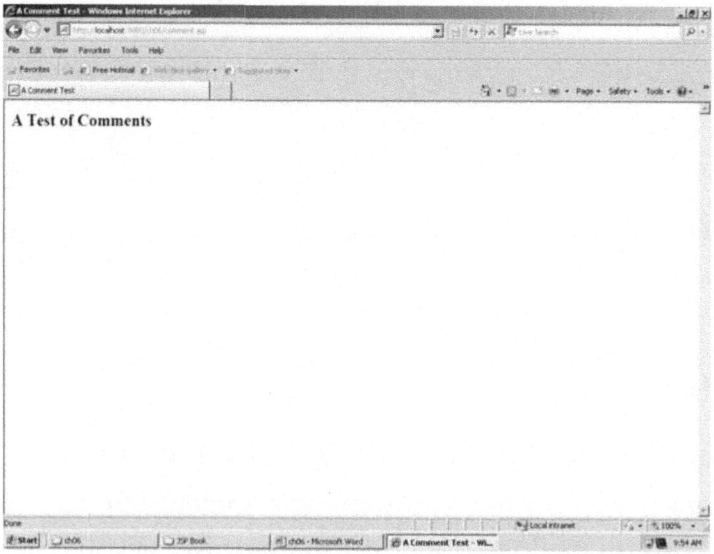

Figure 6.1 Output of comment.jsp

6.3 DECLARATION

Declares a variable or method valid in the scripting language used in the page. A declaration declares one or more variables or methods that you can use in Java code later in the JSP page. You must declare the variable or method before you use it in the JSP page.

You can declare any number of variables or methods within one declaration element, as long as you end each declaration with a semicolon. The declaration must be valid in the Java programming language.

The JSP you write turns into a class definition. All the scriptlets you write are placed inside a single method of this class.

When you write a declaration in a JSP page, remember these rules.

- You must end the declaration with a semicolon (the same rule as for a Scriptlet, but the opposite of an Expression).
- You can use variables or methods that are already declared in packages imported by the page directive, without declaring them in a declaration element.

A declaration has translation unit scope, so it is valid in the JSP page and any of its static include files. A static include file becomes a part of the source of the JSP page and is any file included in an include directive or a static resource included with a <jsp:include> element. The scope of a declaration does not include dynamic resources included with <jsp:include>.

JSP syntax

```
<%! declaration; [ declaration; ]+ ... %>
```

OR

```
<jsp:declaration>
  code fragment [ declaration; ]+ ...
</jsp:declaration>
```

XML syntax

```
<jsp:declaration>
  code fragment [ declaration; ]+ ...
</jsp:declaration>
```

6.3.1 Variable Declaration

A variable declaration is defined as declaration of the servlet class into which the JSP page is translated and compiled. Consider the following declaration of three variables.

```
<%! private int x=0,y=0; private String units="ft"; %>
```

This declaration will have the effect of creating three instance variables in the servlet created for the jsp, named x,y and z units. When declaring JSP instance variable, it is important to keep in mind the potential that multiple thread will be accessing a JSP simultaneously, representing multiple simultaneous page requests.

6.3.2 Method Declaration

Method defined via declarations becomes method of the servlet class into which the JSP page is compiled. For example, the following declaration defines a method for computing factorials.

***Example* 6.2**

Factorial finding using method declaration.

```
<% ! public long fact (long x)
{
if(x= =0 )return 1;
else
return x*fact(x–1);
}
 %>
```

Declaration method can be accessed by all scripting elements on the page, regardless of the order in which the method declaration occurs relative to other scripting elements. In addition, multiple method definition can appear within a single declaration tag, like combination of both variables and method declarations.

***Example* 6.3** declaration.jsp

This example create a method getDate() and display date using the following codes

```
<%@ page import="java.util.*" %>
<html>
<body>
<%!
   date thedate = new date();
   date getdate()
   {
      system.out.println( "in getdate() method" );
      return thedate;
   }
%>
hello! the time is now <%= getdate() %>
</body>
</html>
```

The output of declaration. jsp is shown in Figure 6.2.

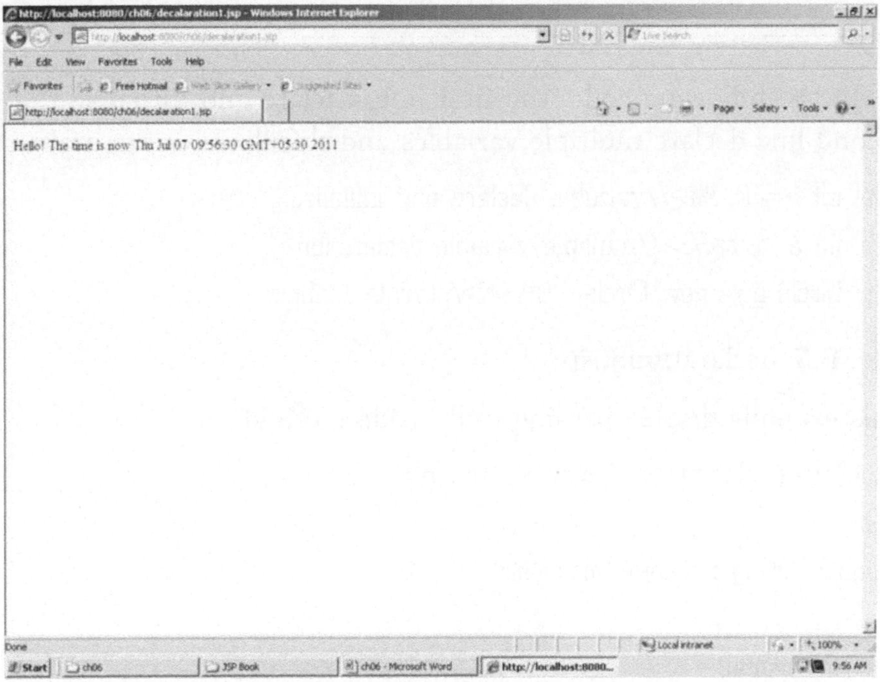

Figure 6.2 Output of declaration.jsp

The example has been created a little contrived, to show variable and method declarations.

Here we are declaring a Date variable theDate, and the method getDate. Both of these are available now in our scriptlets and expressions.

But this example no longer works. The date will be the same, no matter how often you reload the page. This is because these are declarations, and will only be evaluated once when the page is loaded (Just as if you were creating a class and had variable initialization declared in it.).

***Example* 6.4** Returning a string

```
<%! int i; %>
<%! int i=0; %>
<%! public String methodZ(int a)
{
```

```
return "some string"
}
% >
```

In the following code, the first line, declare and initialize the variables, second line declare multiple variables and thirdline create a object.

```
<%! int i = 0; %>//variable declare and initialize.
<%! int a, b, c; %>//multiple variable declaration
<%! Circle a = new Circle(4.0); %>//create a object
```

***Example* 6.5** declaration3.jsp

This example display passing null inside method

```
<%@ page  language="java" errorPage="" %>
<%!
  public String nullconv(String str)
  {
   if(str= =null)
   str="";
   else if(str.equals("null"))
 str="";
 else if((str.trim()).equals(""))
 str="";
 else if(str.equals(null))
 str="";
else
 str=str.trim();
 return str;
 }
%>
<%
 String myVariable=null;
%>
```

```
<html>
<head>
<title>Declaration in JSP</title>
</head>
<body>
This is null variable : <%=nullconv(myVariable)%>
</body>
</html>
```

The output of declaration3.jsp is shown in Figure 6.3.

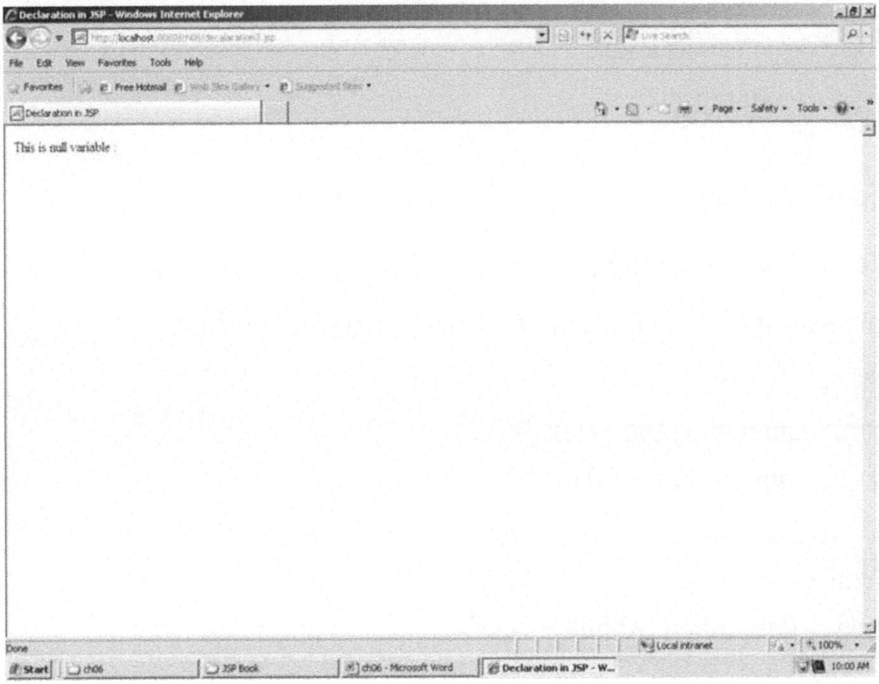

Figure 6.3 Output of declaration3.jsp

***Example* 6.6** declaration 4.jsp

This example display passing null inside method using XML

```
<jsp:declaration>
        public String nullconv(String str)
```

```
        {
          if(str= =null)
            str="";
          else if(str.equals("null"))
            str="";
          else if((str.trim()).equals(""))
            str="";
          else if(str.equals(null))
            str="";
          else
            str=str.trim();
          return str;
        }
</jsp:declaration>
```

***Example* 6.7** declaration5.jsp

This example display name and date without method

```
<%!
 string name = "ANKUR SAXENA";
 string date = "5 JULY, 2011";
%>
<html>
 <title>Declaration Example</title>
 <body>
  this page was last modified on <%= date %> by <%= name %>.
 </body>
</html>
```

The output of declaration5.jsp is shown in Figure 6.4.

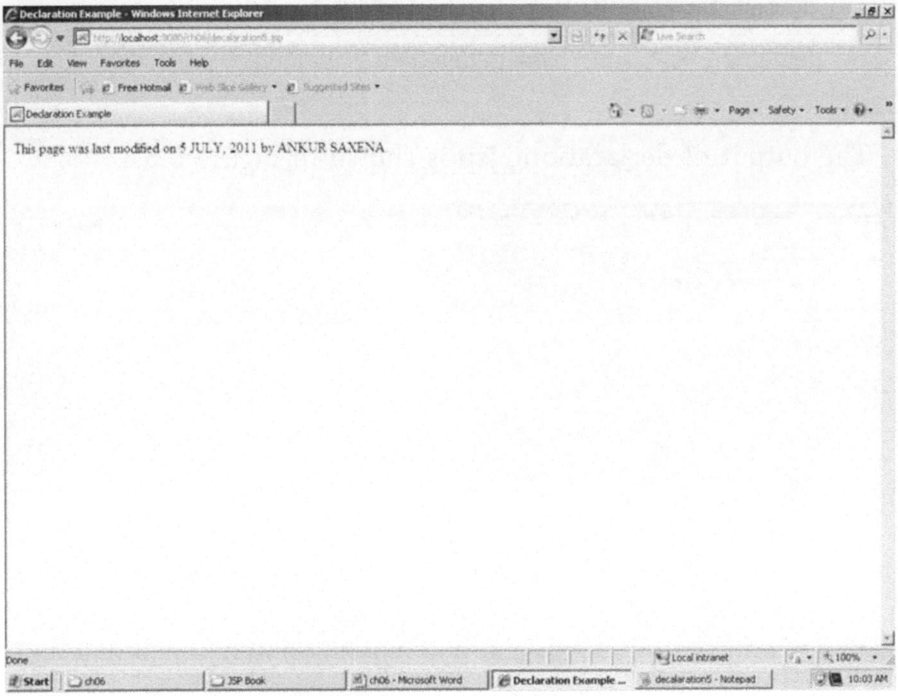

Figure 6.4 Output of declaration5.jsp

***Example* 6.8** declaration6.jsp

This example display name and date using get method.

```
<%!
 private string getname()
 {2
  return "SAXENA";
 }
private int getage()
 {
  return 30;
 }
%>
<html>
```

```
  <head> <title>declaration tag - methods</title> </head>
  <%= getname() %>, age <%= getage() %>, is good!
</html>
```

The output of declaration6.jsp is shown in Figure 6.5.

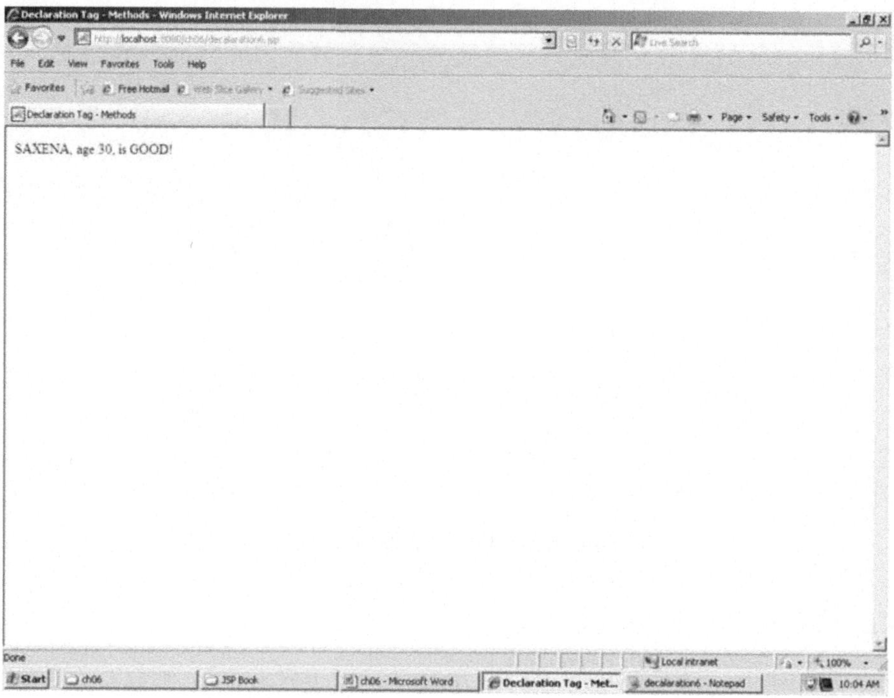

Figure 6.5 Output of declaration6.jsp

***Example* 6.9** declaration 7.jsp

This example display name with passing the parameter

```
<%!
   int a=3;
   public string abc(string name)
   {
     return "hello," +name+"!";
   }
%>
```

```
<html>
<body>
<%=a %>
<%= abc("ANKUR SAXENA") %>
</body>
</html>
```

The output of declaration7.jsp is shown in Figure 6.6.

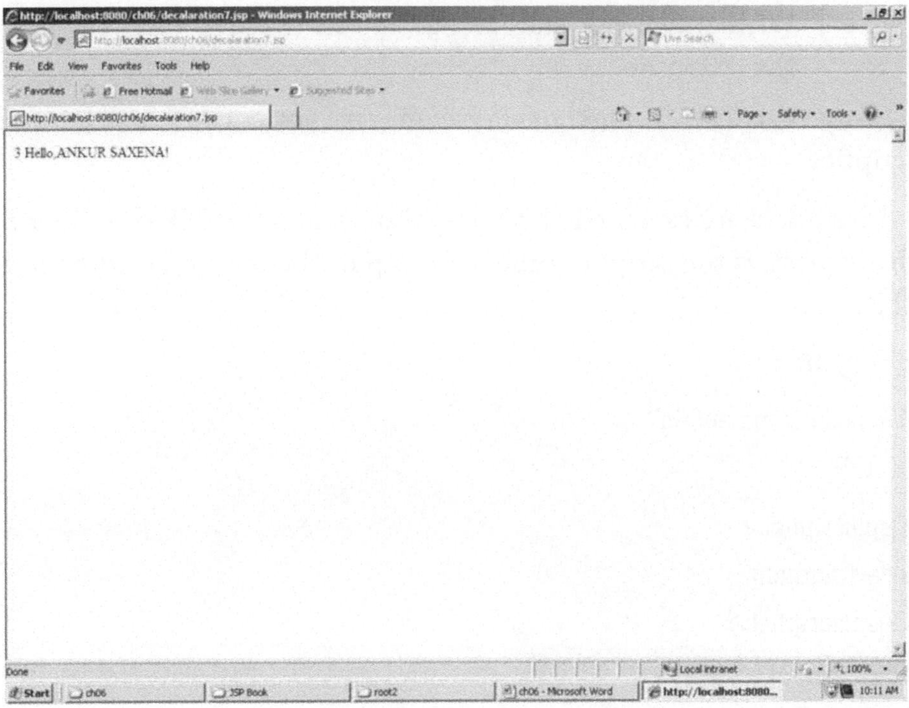

Figure 6.6 Output of declaration7.jsp

6.4 SCRIPTLET

A scriptlet can contain any number of language statements, variable or method declarations, or expressions that are valid in the page scripting language. When the scripting language is set to java, a scriptlet is transformed into a Java programming language statement fragment and is inserted into the service method of the JSP page's servlet. A programming language variable created within a scriptlet is accessible from anywhere within the JSP page.

We can embed any amount of java code in the JSP Scriptlets. JSP Engine places these codes in the jspService() method.

Within a scriptlet, you can do any of the following.

- Declare variables or methods to use later in the JSP page.
- Write expressions valid in the page scripting language.
- Use any of the implicit objects or any object declared with a <jsp:useBean> element.
- Write any other statement valid in the scripting language used in the JSP page.

Any text, HTML tags, or JSP elements you write must be outside the scriptlet.

Scriptlets are executed at request time, when the JSP container processes the request. If the scriptlet produces output, the output is stored in the "out" object.

JSP syntax

```
<% code fragment %>
    OR
<jsp:scriptlet>
code fragment
</jsp:scriptlet>
```

XML syntax

```
<jsp:scriptlet>
 code fragment
</jsp:scriptlet>
```

***Example* 6.10** loop.jsp

This example display dimonstration of loop using scriplets.

```
<html>
<body>
<%
```

```
for(int i=1;i<=20;i++)
{
out.println("hello Ankur");
}
%>
</body>
</html>
```

The output of loop.jsp is shown in Figure 6.7.

Figure 6.7 Output of loop.jsp

Example 6.11 tab.jsp

This example display table of a given number.

```
<html>
<body>
<%! int a=1,b=4,c; %>
<%
```

```
    while(a<=10)
    {
        c=b*a;
      out.println(c); %>
<br>
      <% a++;
    }
%>
</body>
</html>
```

The output of tab.jsp is shown in Figure 6.8.

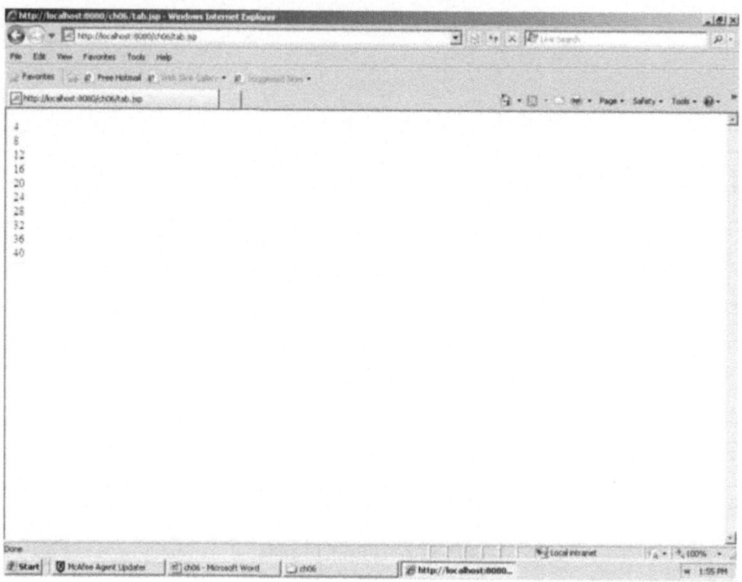

Figure 6.8 Output of tab.jsp

***Example* 6.12** number.jsp

This example display numbers 1 to 5.

```
<%
        for(int i=1; i<6; i++) {
  %>
```

```
        <div> <%= i %> </div>
    <%}%>
```

The output of number.jsp is shown in Figure 6.9.

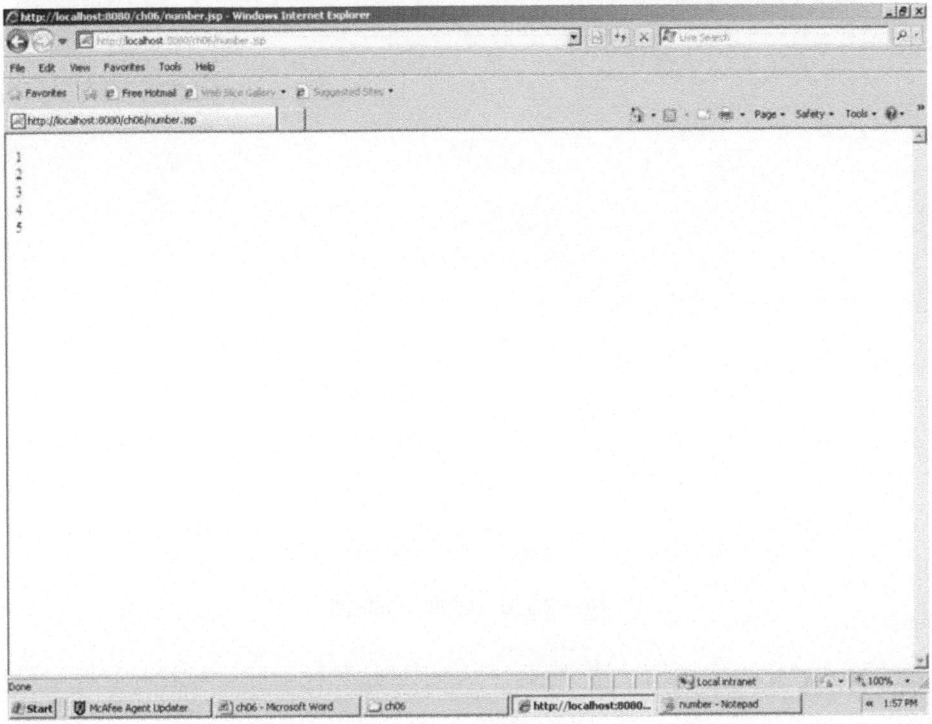

Figure 6.9 Output of number.jsp

***Example* 6.13** date.jsp

This example display current date and time using java utilities.

```
<html>
<body>
<%
   system.out.println( "Evaluating date now" );
   java.util.date date = new java.util.Date();
%>
hello!  the time is now <%= date %>
```

```
</body>
</html>
```

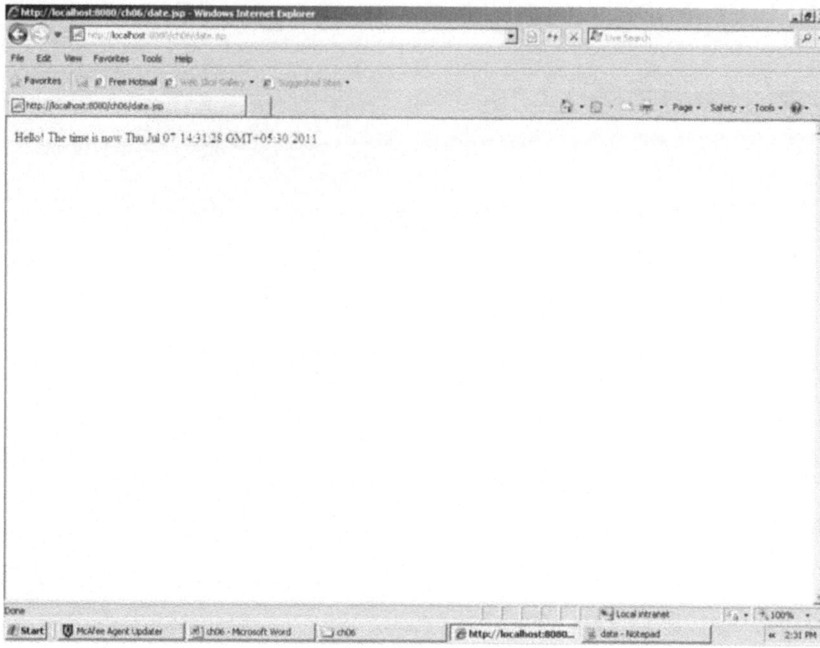

Figure 6.10 Output of date.jsp

***Example* 6.14** date1.jsp

This example display current date, time and machine address.

```
<html>
<body>
<%
   // this scriptlet declares and initializes "date"
   system.out.println( "Evaluating date now" );
   java.util.date date = new java.util.Date();
%>
hello!  the time is now
<%
   out.println( date );
   out.println( "<BR>your machine's address is " );
```

```
    out.println( request.getRemoteHost());
%>
</body>
</html>
```

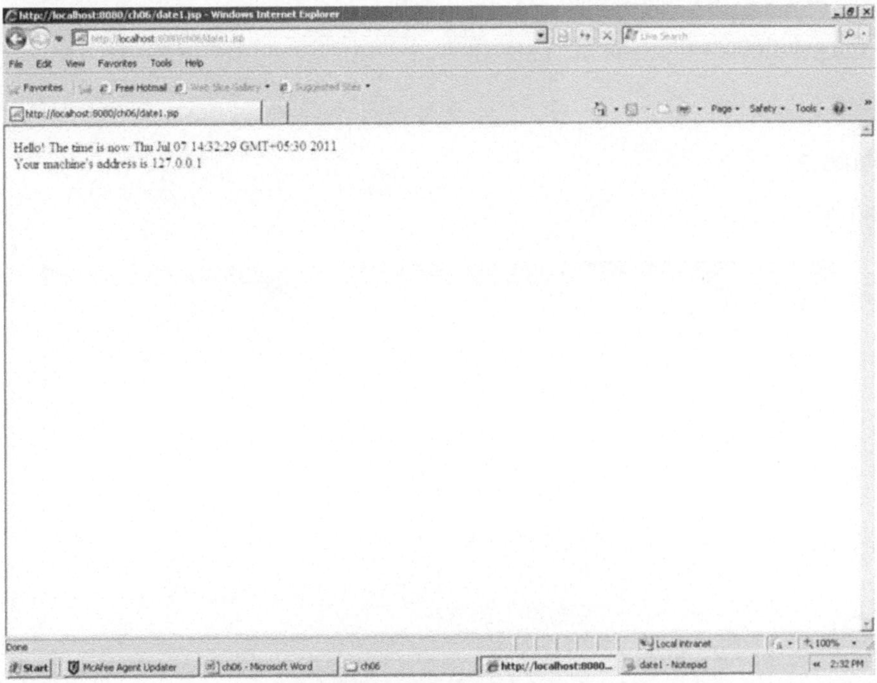

Figure 6.11 Output of date1.jsp

***Example* 6.15** date2.jsp

This example display how to declare and initialize date and scriplets to generate HTML ouptput.

```
<html>
<body>
<%
    // this scriptlet declares and initializes "date"
    system.out.println( "Evaluating date now" );
    java.util.date date = new java.util.Date();
```

```
%>
hello!  the time is now
<%
    // this scriptlet generates HTML output
    out.println( string.valueof( date ));
%>
</body>
</html>
```

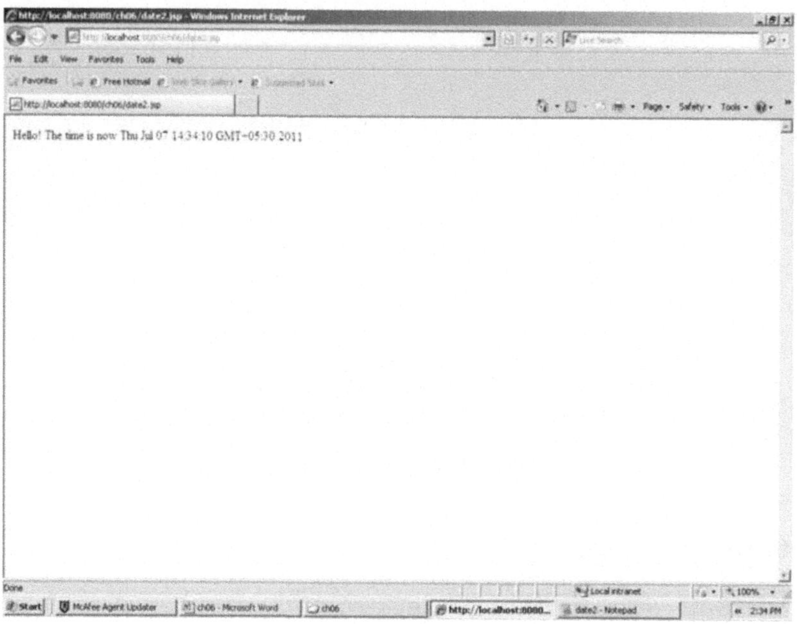

Figure 6.12 Output of date2.jsp

This code shows how to create a package foo and set all setter and getter method inside this package

```
<%
  String name = null;
  if (request.getParameter("name") = = null) {
%>
<%@ include file="error.html" %>
```

```
<%
  } else {
  foo.setName(request.getParameter("name"));
  if (foo.getName().equalsIgnoreCase("integra"))
    name = "acura";
  if (name.equalsIgnoreCase( "acura" )) {
%>
```

Examples 6.16 mes.jsp

This example show JSP scriptlet is embedded in HTML using a similar tag structure, but without the equals sign.

```
<html>
 <body>
   <%
     //Set a variable and then output it using a shorthand JSP expression.
     String outputMessage = "Content printed from a JSP expression.";
   %>
   <%= outputMessage %> <br />
 </body>
</html>
```

The output of mes.jsp is shown in Figure 6.13.

Examples 6.17 fib.jsp

Once again a more complicated example shows how JSP scriptlets can be used to generate dynamic output using JSP expressions.

```
<html>
 <body>
  <%
    String fibonacciOutput = "";
    int lastInt = 0;
    int currentInt = 1;
```

Figure 6.13 Output of mes.jsp

The output of fib5.jsp is shown in Figure 6.14.

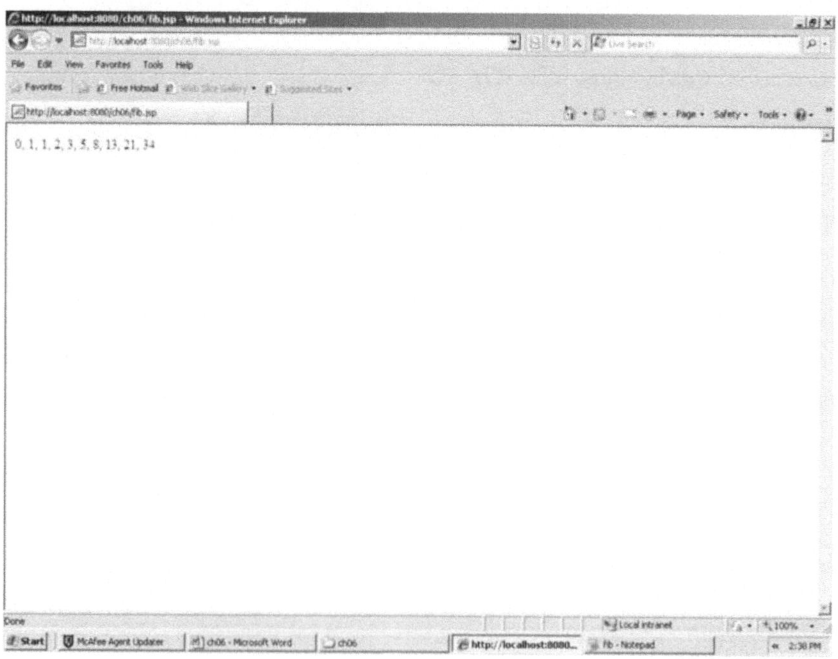

Figure 6.14 Output of fib5.jsp

```
  int nextInt = 1;
    int iterations = 10;
for(int counter = 0; counter < iterations; counter++)
    {
      if(counter + 1 < iterations)
      {
        fibonacciOutput += lastInt + ", ";
      }
      else
      {
        fibonacciOutput += lastInt;
      }
      lastInt = currentInt;
      currentInt = nextInt;
      nextInt = currentInt + lastInt;
    }
  %>
  <%= fibonacciOutput %> <br />
 </body>
</html>
```

Examples 6.18 temp.jsp

This example display temperature in fahrenheit and celsius.

```
<%@ page import="java.text.*" session="false"%>
<html>
<head>
<title>Scriptlet Example</title>
</head>
<body>
<table border="0" cellpadding="3">
<tr>
```

```
    <th>Fahrenheit</th>
    <th>Celsius</th>
  </tr>
  <%
    NumberFormat fmt = new DecimalFormat("###.000");
    for (int f = 32; f <= 212; f += 20) {
      double c = ((f-32) * 5) / 9.0;
      String cs = fmt.format(c);
  %>
    <tr>
      <td align="RIGHT"> <%= f %> </td>
      <td align="RIGHT"> <%= cs %> </td>
    </tr>
  <%
    }
  %>
  </table>
  </body>
  </html>
```

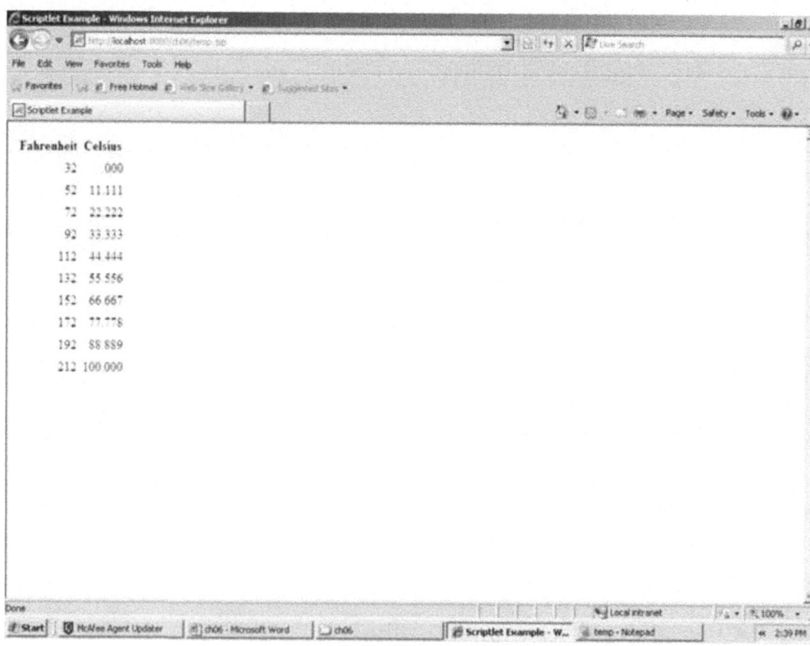

Figure 6.15 Output of temp.jsp

The output of temp.jsp is shown in Figure 6.15.

Examples 6.19 *avg.jsp*

This example display average grade.

```
<html>
 <head>
  <title>finding average grades</title>
 </head>
 <body>
  <h1>finding average grades</h1>
  <%
    double grades[] = {88, 98, 73, 56, 86, 64};
    double sum, average;
    sum = 0;
    for (int i = 0; i < grades.length; i++) {
      sum += grades[i];
    }
    average = sum / grades.length;
    out.println("average grade = " + average);
  %>
 </body>
</html>
```

The output of avg.jsp is shown in Figure 6.16.

Examples 6.20 *new.jsp*

This example show casting a new type.

```
<html>
 <head>
  <title>casting to a new type</title>
 </head>
 <body>
  <h1>casting to a new type</h1>
```

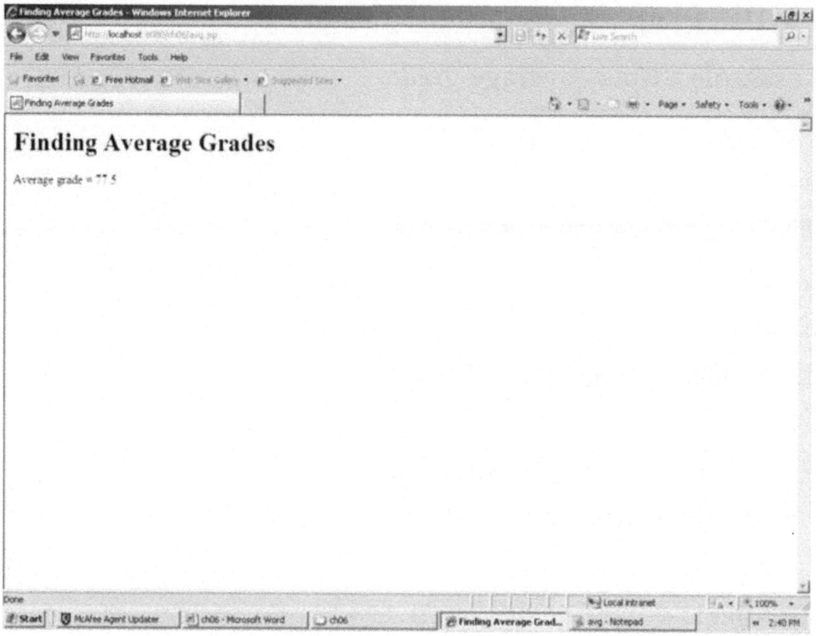

Figure 6.16 Output of avg.jsp

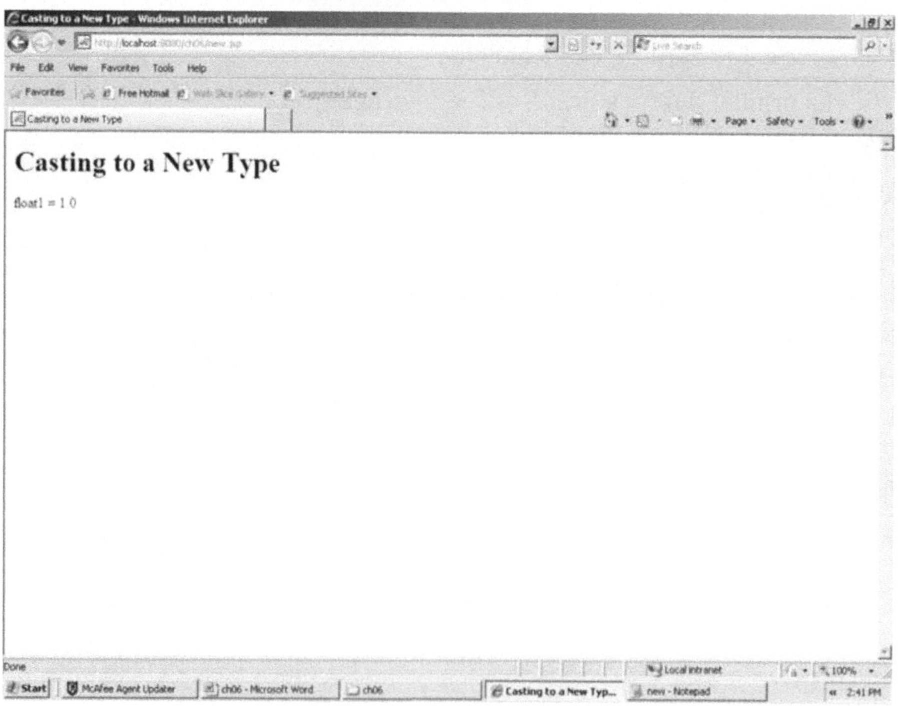

Figure 6.17 Output of new.jsp

```
    <%
        float float1;
        double double1 = 1;
        float1 = (float) double1;
        out.println("float1 = " + float1);
    %>
  </body>
</html>
```

The output of new.jsp is shown in Figure 6.17.

6.5 MIXING SCRIPTLETS AND HTML

We have already seen how to use the "out" variable to generate HTML output from within a scriptlet. For more complicated HTML, using the out variable all the time loses some of the advantages of JSP programming. It is simpler to mix scriptlets and HTML.

Suppose you have to generate a table in HTML then this is a common operation, and you may want to generate a table from a SQL table, or from the lines of a file. But to keep our example simple, we will generate a table containing the numbers from 1 to N, just to show the technique.

```
<TABLE BORDER=2>
<%
   for ( int i = 0; i < n; i++ ) {
      %>
      < tr >
      <td>Number</td>
      <td> <%= i+1 %> </td>
      </tr>
      <%
   }
%>
</TABLE>
```

You would have to supply an int variable "n" before it will work, and then it will output a simple table with "n" rows.

The important things to notice are how the %> and <% characters appear in the middle of the "for" loop, to let you drop back into HTML and then to come back to the scriptlet.

The concepts are simple here. As you can see, you can drop out of the scriptlets, write normal HTML, and get back into the scriptlet. Any control expressions such as a "while" or a "for" loop or an "if" expression will control the HTML also. If the HTML is inside a loop, it will be omitted once for each iteration of the loop.

Another example of mixing scriptlets and HTML is shown below. Here it is assumed that there is a boolean variable named "hello", available. If you set it to true, you will see one output, if you set it to false, you will see another output.

```
<%
  if ( hello ) {
     %>
     < P >Hello, Ankur
     <%
  } else {
     %>
     < P>Goodbye, Ankur
     <%
  }
%>
```

It is a little difficult to keep track of all open braces, and scriptlet starts and ends, but with a little practice and some good formatting discipline, you will acquire competence in doing it.

6.6 EXPRESSION

A JSP expression is used to insert the value of a scripting language expression, converted into a string, into the data stream returned to the client. When the scripting language is the Java programming language, an expression is transformed into a statement that converts the value of the expression into a String object and inserts it into the implicit "out" object.

The expression element can contain any expression that is a valid page scripting language. When the Java programming language is the scripting language you do not use a semicolon to end the expression. However, the same expression within a scriptlet requires the semicolon.

You can sometimes use expressions as attribute values in JSP elements. An expression can be complex and composed of more than one part or expression. The parts of an expression are evaluated in left-to-right order.

JSP expressions are designed to output the value of a single Java variable or expression. They can be linked to Javascript's eval() function except that a JSP expression outputs the result into the content of the HTML file.

```
<html>
 <body>
   <%= 5+5 %>
 </body>
</html>
```

This extremely abbreviated example demonstrates a JSP expression that has been inserted into an HTML page. The contents of the expression, which is in this case the math problem "5+5" is evaluated at runtime and inserted into the HTML that is delivered to the client side. So the web user who views the JSP webpage will see.

```
<html>
 <body>
   10
 </body>
</html>
```

Obviously this is a very contrived and rather pointless example. However the true power of JSP becomes more obvious when you output the value of variables or class methods.

This code display date using expression

```
<html>
 <body>
    <%= (new java.util.Date()) %>
 </body>
</html>
```

JSP syntax

```
<%= expression %>
   OR
<jsp:expression>
  expression
</jsp:expression>
```

XML syntax

```
<jsp:expression>
  expression
</jsp:expression>
```

***Example* 6.21** exp1.jsp

This example display some expression results

```
<html>
  <head>
  <title>JSP Expression Page</title>
 </head>
 <body>
<h2>JSP Expression Page</h2>
<p> <%= 3+3 %> <br>
<%= 2*3 %> <br>
<%= (2*3) + (3+3) %> <br>
<%= "Here is some text " + (2*3) + (3+3) %> <br>
<%= "Here is some text " + ((2*3) + (3+3)) %> <br>
<%! String aString = "What is two times 3? "; %>
```

```
<%! int x = 2; %>
<%! int y = 3; %>
<%= aString + (x*y) %>
</p>
</body>
 </html>
```

The output of exp1.jsp is shown in Figure 6.18.

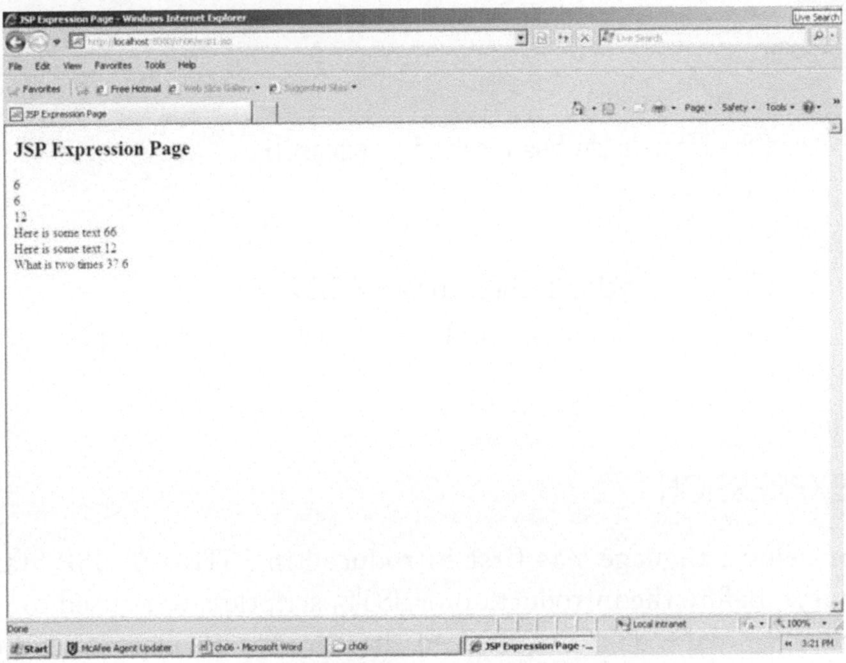

Figure 6.18 Output of exp1.jsp

Example 6.22

Let us see an example of a JSP page with declarations, scriptlet and expression.

```
<%@ page language="java" contenttype="text/html; charset=utf-8"
    pageencoding="utf-8"%>
  <%!
  //this is declaration
  int counter;
  public void jspinit() {
```

```
        counter = 0;
    }
    %>
<html>
<head>
<title>jsp scripting elements example</title>
</head>
<body>
        <%
        //this is a scriptlet
        log("the lifecycle jsp has received a request");
        counter++;
        %>
        <h2>jsp life cycle : request counter</h2>
        <p>this page has been called <%=counter %> times </p>
    </body>
</html>
```

6.7 EL EXPRESSION

Expression Language was first introduced in JSTL 1.0 (JSP Standard Tag Library). Before the introduction of JSTL, scriptlets were used to manipulate application data. JSTL introduced the concept of an expression language (EL) which simplified the page development by providing standard tag libraries. These tag libraries provide support for common, structural tasks, such as, iteration and conditionals, processing XML documents, internationalization and database access using the Structured Query Language (SQL).

Contains an expression in the JSP Expression Language (EL). It allows easy access to application data stored in JavaBeans components.

A primary feature of JSP technology is its support for an expression language (EL). An expression language makes it possible to easily access application data stored in JavaBeans components.

Some action elements have attributes whose value can be computed at request time. In JSP syntax, the format of such a value is the same as a JSP expression: <%= expression %>. In XML syntax, the format of the value is %= expression %. When using the JSP expression language (EL), the format of the value in either syntax is ${ expression }.

EL expressions can be quoted by adding a backslash to the front of the expression when the expression language is enabled. This prevents the expression from being evaluated. Example: \${x+y} will display as ${x+y}.

All elements in XML syntax can contain an xmlns attribute that complies with the XML namespace spec. This attribute is used to declare tag libraries and other namespaces. Within the start tag of an element, its syntax is [xmlns:taglibprefix="URI"]+.

All JSP standard actions and custom actions can contain a jsp:attribute standard element as a substitute for any of its attributes. The jsp:attribute syntax is <jsp:attribute name="attributeName" [trim="true | false"] />

If an action contains any jsp:attribute elements and the action also has a body, it must use the jsp:body tag to represent the body. The jsp:body syntax is <jsp:body> any elements or text </jsp:body>

EL expressions can be used

- in static text.
- in any standard or custom tag attribute that can accept an expression.

The value of an expression in static text is computed and inserted into the current output. If the static text appears in a tag body, note that an expression will not be evaluated if the body is declared to be tag dependent (body-content Attribute).

body-content Attribute

You specify the type of a tag's body content using the body-content attribute:

bodycontent="empty | scriptless | tagdependent"

You must declare the body content of tags that do not accept a body as empty. For tags that have a body there are two options. Body content containing custom and standard tags and HTML text is specified as scriptless. All other types of body content, for example, SQL statements passed to the query tag is specified as tagdependent. If no attribute is specified, the default is scriptless.

There are three ways to set a tag attribute value.

1. *With a single expression construct* <some:tag value="${expr}"/> The expression is evaluated and the result is coerced to the attribute's expected type.

2. *With one or more expressions separated or surrounded by text* <some:tag value="some${expr}${expr}text${expr}"/> The expressions are evaluated from left to right. Each expression is coerced to a String and then concatenated with any intervening text. The resulting String is then coerced to the attribute's expected type.

3. *With text only* <some:tag value="sometext"/> In this case, the attribute's String value is coerced to the attribute's expected type.

Syntax (JSP and XML)

```
${Expression}
Expression = {(ChoiceExpression | (Expression BinaryOp Expression)
             (UnaryOp Expression) | Value }
ChoiceExpression = Expression ? Expression : Expression
BinaryOp (JSP syntax) = and | && | or | || | '+' | '-' | * | / | did
                      | % | mod | > | gt | < | lt | >= | ge | <= |
                   le | == | eq | != | ne
BinaryOp (XML syntax) = and | && | or | || | '+' | '-' | * | / | div
                   | % | mod | gt | lt | ge | le | == | eq | ne
UnaryOp = {'-' | ! | not | empty }
Value = {ValuePrefix | (Value ValueSuffix) }
ValuePrefix = {Literal | '('Expression')' | ImplicitObject |
     Java language identifier | FuncInvocation }
```

ValueSuffix = { . Identifier | '['Expression']'}

FuncInvocation = [Identifier :]Identifier '('[Expression [',' Expression]+]')'

Literal = {true|false} | (0-9)+ | FloatingPtLiteral | StringLiteral | null

FloatingPtLiteral = [0-9]+ . (0-9)+ [Exponent] | (0-9)+ [Exponent]

Exponent = {e | E} ['+' | -] (0-9)+

StringLiteral = {'[^(' | \) | \' | \\]+' | "[^(" | \) | \" | \\]+"}

6.7.1 Attributes

- ' as \'. This is required within a single quote-delimited attribute value.
- " as \". This is required within a double quote-delimited attribute value.
- \ as \\
- %> as %\>
- <% as <\%
- ' and " can be used to indicate single and double quotes.

6.7.2 Typographical Conventions

code = fixed

bold = default

italics = user-defined

| = or

[ ] = optional

{ } = required choice

... = list of items

+ = can repeat

' ' = literal

() = grouped together

^ = anything other than

– = a range

6.7.3 Variables

The Web container evaluates a variable that appears in an expression by looking up its value according to the behaviour of PageContext.findAttribute(String). For example, when evaluating the expression ${product}, the container will look for product in the page, request, session, and application scopes and will return its value. If product is not found, null is returned. A variable that matches one of the implicit objects will return that implicit object instead of the variables value.

Properties of variables are accessed using the dot (.) operator and can be nested arbitrarily.

The JSP expression language unifies the treatment of the . and [] operators. expr-a.identifier-b is equivalent to expr-a["identifier-b"]; that is, the identifier, identifier-b is used to construct a literal whose value is the identifier, and then the []operator is used with that value.

To evaluate expr-a[expr-b], evaluate expr-a into value-a and evaluate expr-b into value-b. If either value-a or value-b is null, return null.

- If value-a is a Map, return value-a.get(value-b). If !value-a.containsKey(value-b), then return null.
- If value-a is a List or array, coerce value-b to int and return value-a.get(value-b) or Array.get(value-a, value-b), as appropriate. If the coercion could not be performed, an error is returned. If the call returns an Indexout of Bounds Exception, null is returned. If the call returns another exception, an error is returned.
- If value-a is a JavaBeans object, coerce value-b to String. If value-b is a readable property of value-a, then return the result of a get call. If the get method throws an exception, an error is returned. A variable that matches one of the implicit objects will return that implicit object instead of the variable's value.

6.7.4 Implicit Objects

The JSP expression language defines a set of implicit objects.

pagecontext The context for the JSP page, which Provides access to various objects are given below.

servletContext The context for the JSP page's servlet and any web components contained in the same application.

session The session object for the client.

request The request triggering the execution of the JSP page.

response The response returned by the JSP page.

In addition, several implicit objects are available that allow easy access to the following objects.

param Maps a request parameter name to a single value

paramValues Maps a request parameter name to an array of values

header Maps a request header name to a single value

headerValues Maps a request header name to an array of values

cookie Maps a cookie name to a single cookie

initParam Maps a context initialization parameter name to a single value

Finally, there are objects that allow access to the various scoped variables described in using scope objects.

pageScope Maps page-scoped variable names to their values

requestScope Maps request-scoped variable names to their values

sessionScope Maps session-scoped variable names to their values

applicationScope Maps application-scoped variable names to their values

When an expression references one of these objects by name, the appropriate object is returned instead of the corresponding attribute. For example, `${pageContext}` returns the PageContext object, even if there is an existing pageContext attribute containing some other value.

Note EL implicit objects are not as same as the implicit objects available for JSP scripting except for pageContext.

JSP and EL implicit objects have only one object in common (pageContext), and pageContext has properties for accessing all of the other eight JSP implicit objects.

The first four maps represent the various attribute scopes which can be used to look up identifiers in specific scopes. The next four maps are for fetching the values of request parameters and headers.The cookie implicit object provides access to the cookies set by a request and the final EL implicit object, initParam, is a map storing the names and values of any context initialization parameters associated with the web application.

Let us look at few examples on how to use implicit objects,

Example 6.23

This example show the application of param implicit object.

First.jsp

```
First Name:
  <input type='text' name='Name'/>
 Last Name:
  <input type='text' name='Address'/>
  <input type='submit' value='Submit'/>
```

Second.jsp

```
Name is : ${param.Name}
Address is : ${param.Address}
```

Similarly,

```
Header Key : ${header.key}
Header Value : ${header.value}
```

Here in First.jsp we are entering the details of Name and Address, when we click on submit button the flow goes to Second.jsp. We can retrieve the values by using ${param.xxxx}.

Scope examples

Accessing a page-scoped attribute named as "firstName"

```
${pageScope.firstName}
```

Accessing a request-scoped attribute named as "firstName"

```
${requestScope.firstName}
```

Accessing a session-scoped attribute named as "lastName" (null if not found)

```
${sessionScope.lastName}
```

Page context examples

```
${pageContext.request.protocol}
${pageContext.response.locale}
${pageContext.session.id}
${pageContext.servletContext}
```

In the above we are accessing request, response, session, and application properties, using the pageContext implicit object.

***Example* 6.24**

This example show the application of cookie implicit object.

Cookie example

addCookies.java

```
Cookie c=new Cookie("name","test");
res.addCookie(c);
Test.jsp
Name  : ${cookie["name"]}
Value : ${cookie["name"].value}
```

Herein addCookies.java, we are creating a new Cookie,whose values can be retrieved in jsp by using ${cookie.xxxx}.

***Example* 6.25** Init parameter example

This example show the application of initparam implicit object

XML

```
<context-param>
      <param-name>email</param-name>
      <param-value>Hello@xyz.com</param-value>
</context-param>
```

Test.jsp

```
email is: ${initParam.email
```

6.7.5 Literals

A literal represent a value of a certain type where the type describe the behaviour of the value.

Boolean Boolean literals are simple. There are only two logical values that a boolean value can have, true and false. The values of true and false do not convert into any numerical representation.

Interger Intergers are most commonly used type in the typical program. There are several interger literal such as int, long, as in the Java programming language.

Floating point Floating-point numbers represent decimal values with a fractional component. They can be expressed in either standard or scientific notation.

String with single and double quotes; "is escaped as\", is escaped as \; and \ is escaped as \\.

Null Null the null literal can be assigned to any object reference.

6.7.6 Operators

In addition to the dot (.) and [] operators discussed in variables, the JSP expression language provides the following operators.

Operator	Description
.	Access a bean property or Map entry.
[]	Access an array or List element.
()	Grouping of a subexpression for evaluation order.
? :	Conditional test: condition ? if True : if False.
+	Addition.
-	Subtraction.
*	Multiplication.
/or div	Division.
% or mod M	Modulo (remainder).

== or eq	Equality.
!= or ne	Inequality.
< or lt	Less than.
> or gt	Greater than.
<= or le	Less than or equal.
>= or ge	Greater than or equal.
&& or and	Logical AND.
|| or or	Logical OR.
! or not	Unary Boolean complement.
empty	Test for null,empty String,array or Collection.
func(args)	A function call.

Arithmetic operators

Before JSP2.0

This code display application of arithmetic operator.

```
<%!
        int k = 0;
%>
<%
        k = 10 + 20;
        out.println("value of k is.."+k);
%>
```

Using JSP2.0/JSTL (without using scriptlet)

```
value of k is ${10 + 20}
```

Few more examples on arithmetic operations.

EL expression	**Result**
${10.2 + 20.3}	O/P: 30.5
${–40 – 20}	O/P: –60

${20 * 2}	O/P: 40
${3/4}	O/P: 0.75
${3 div 4}	O/P: 0.75
${10/0}	O/P: Infinity
${50%8}	O/P: 2
${50 mod 8}	O/P: 2
${(10==20) ? “true” : “false”}	O/P: false

Comparison operators

Numeric

EL expression	**Result**
${10 < 20}	true
${10 lt 20}	true
${1 > (4/2)}	false
${1 gt (4/2)}	false
${4.0 >= 3}	true
${4.0 ge 3}	true
${40 <= 30}	false
${40 le 30}	false
${10 == 10}	true
${100 eq 100}	true
${(5*20) != 100}	false
${(5*20) ne 100}	false

Alphabetic

EL expression	**Result**
${‘a’ < ‘b’}	true
${‘4’ > 3}	true

Logical and, (&&) or, (| |) not, (!)

Relational Comparisons can be made against other values, or against boolean, string, integer, or floating point literals.

==, eq, !=, ne, <, lt, >, gt, <=, ge, >=, le.

Empty The empty operator is a prefix operation that can be used to determine whether a value is null or empty.

Conditional A ? B : C. Evaluate B or C, depending on the result of the evaluation of A.

The precedence of operators highest to lowest, left to right is as follows.

- [] .
- ()—Used to change the precedence of operators.
- – (unary) not ! empty
- * / div % mod
- + – (binary)
- < > <= >= lt gt le ge
- == != eq ne
- && and
- | | or
- ? :

6.7.7 Reserved Words

The following words are reserved for the JSP expression language and should not be used as identifiers.

- and
- gt
- instanceof
- ne
- false
- not
- eq
- true
- or
- le
- empty
- lt

- ge
- div
- null
- mod

Note that many of these words are not in the language now, but they may be in the future, so you should avoid using them.

6.7.8 Functions

The JSP Expression Language allows you to define a function that can be invoked in an expression. Functions are defined using the same mechanisms as custom tags.

Functions can appear in static text and tag attribute values.

For example, the following example page imports the /functions library and invokes the function equals in an expression.

```
<%@ taglib prefix="f" uri="/functions"%>
...
   <c:when
     test="${f:equals(selectedLocaleString,
      localeString)}" >
```

To define a function you program it as a public static method in a public class. The mypkg.MyLocales class example below defines a function that tests the equality of two strings as follows.

```
package mypkg;
public class MyLocales {
  ...
  public static boolean equals( String l1, String l2 ) {
    return l1.equals(l2);
  }
}
```

Then you map the function name as used in the EL expression to the defining class and function signature in a Tag library discriptor (TLD). The following functions.tld file maps the equals function to the class containing the implementation of the function equals and the signature of the function.

```
<function>
  <name>equals</name>
  <function-class>mypkg.MyLocales</function-class>
  <function-signature>boolean equals( java.lang.String,
    java.lang.String )</function-signature>
</function>
```

Defining functions The JSP Expression Language allows you to define a function that can be invoked in an expression. Functions are defined using the same mechanisms as custom tags.

Example 6.26

```
public class MyFunctions {
          public static String toCaps( String text ) {
          return text.toUpperCase();
          }
  }
```

Using functions

XML syntax

```
<uri>testFunctionsURI</uri>
     <function>
       <name>toCaps</name>
       <function-class>mypkg.MyFunctions</function-class>
       <function-signature>String toCaps( java.lang.String)</function-signature>
     </function>
```

JSP syntax

```
<%@ taglib prefix="myFunctionTest" uri="testFunctionsURI"%>
     ${myFunctionTest.toCaps("testingCaps")}
```

Example 6.27

The following three examples show tag attributes using EL expressions to access values stored in JavaBeans components.

```
<a:tag value="${x+y}" />
<a:tag value="${first} ${last}" />
<a:tag>${x+y}</a:tag>
```

Table 6.1 Expression language examples

EL Expression	Result
${1 > (4/2)}	false
${4.0 >= 3}	true
${100.0 == 100}	true
${(10*10) ne 100}	false
${'a' < 'b'}	true
${'hip' gt 'hit'}	false
${4 > 3}	true
${1.2E4 + 1.4}	12001.4
${3 div 4}	0.75
${10 mod 4}	2
${!empty param.Add}	True if the request parameter named Add is null or an empty string
${pageContext.request.contextPath}	The context path
${sessionScope.cart.numberOfItems}	The value of the numberOfItems property of the session-scoped attribute named cart
${param['mycom.productId']}	The value of the request parameter named mycom.productId
${header["host"]}	The host
${departments[deptName]}	The value of the entry named deptName in the departments map
${requestScope['javax.servlet.forward.servlet_path']}	The value of the request-scoped attribute named javax.servlet.forwardservlet_path

***Example* 6.28** exl.jsp

```
Expression language example
<html>
<head>
<title>Expression Language Examples</title>
```

```
<%
  // set up a page context parameter for use later in the page
  // normally this would have been set within the context of
  // an application
  pageContext.setAttribute("pageColor", "red");
%>
</head>
<body bgcolor="${pageScope.pageColor}">
<h1>Welcome to the ${param.department} Department</h1>
Here are some basic comparisons:
<p>
Is 1 less than 2? ${1<2} <br>
Does 5 equal 5? ${5= =5} <br>
Is 6 greater than 7? ${6 gt 7}<br>
<p>Now for some math:<br>
6 + 7 = ${6+7}<br>
8 × 9 = ${8*9}<br>
<hr>You appear to be using the following browser:
${header["user-agent"]}
</body>
</html>
```

The output of exl.jsp is shown in Figure 6.19.

***Example* 6.29** *exp3.jsp*

This example display results of EL operators

```
<html>
<body>
   <h1>EL Arithmetic</h1>
   <table border="1">
     <tr>
      <td> <b>Concept</b> </td>
```

```
    <td> <b>EL Expression</b> </td>
    <td> <b>Result</b> </td>
  </tr>
  <tr>
    <td>Literal</td>
    <td>${'${'}10}</td>
```

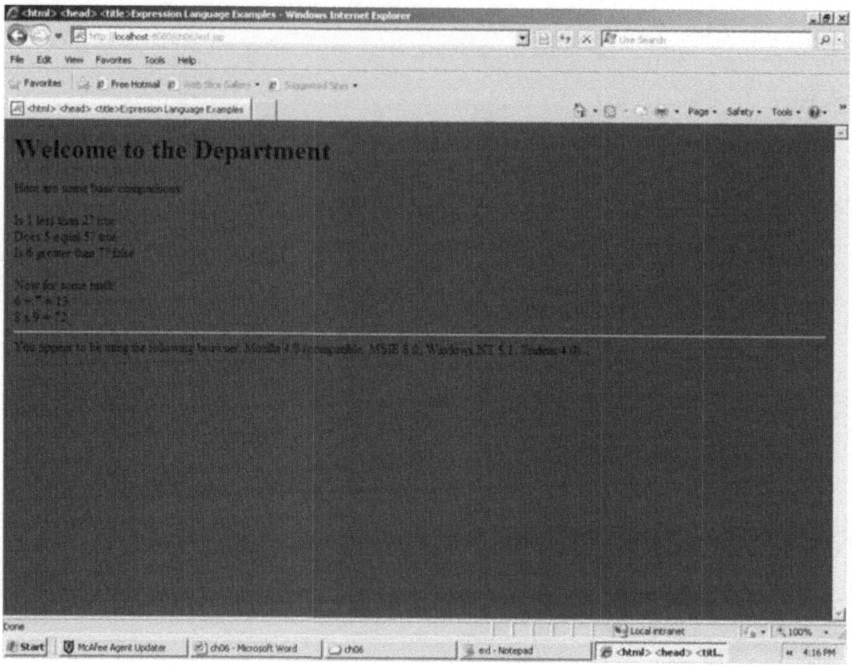

Figure 6.19 Output of exl.jsp

```
    <td>${10}</td>
  </tr>
  <tr>
    <td>Addition</td>
    <td>${'${'}10 + 10 }</td>
    <td>${10 + 10}</td>
  </tr>
  <tr>
    <td>Subtraction</td>
```

```
  <td>${'${'}10 – 10 }</td>
  <td>${10 – 10}</td>
</tr>
<tr>
  <td>Multiplication</td>
  <td>${'${'}10 * 10 }</td>
  <td>${10 * 10}</td>
</tr>
<tr>
  <td>Division / </td>
  <td>${'${'}10 / 3 }</td>
  <td>${10 / 3}</td>
</tr>
<tr>
  <td>Division DIV</td>
  <td>${'${'}10 div 3 }</td>
  <td>${10 div 3}</td>
</tr>
<tr>
  <td>Modulus</td>
  <td>${'${'}10 % 10 }</td>
  <td>${10 % 3}</td>
</tr>
<tr>
  <td>Modulus</td>
  <td>${'${'}10 mod 10 }</td>
  <td>${10 mod 3}</td>
</tr>
<tr>
  <td>Division by Zero</td>
  <td>${'${'}10 / 0 }</td>
```

```
  <td>${10 / 0}</td>
 </tr>
 <tr>
  <td>Exponential</td>
  <td>${'${'}2E2}</td>
  <td>${2E2}</td>
 </tr>
 <tr>
  <td>Unary Minus</td>
  <td>${'${'}-10}</td>
  <td>${-10}</td>
 </tr>
  </table>
 </body>
</html>
```

The output of exp3.jsp is shown in Figure 6.20.

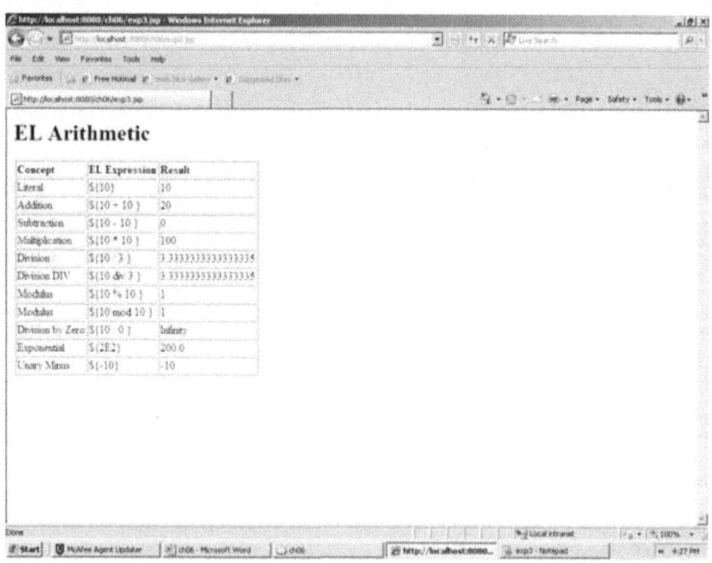

Figure 6.20 Output of exp3.jsp

REVIEW QUESTIONS

1. What is the role of scripting elements in jsp
2. What are the two kinds of comments in JSP and what's the difference between them.
3. How many JSP scripting elements are there and what are they?
4. How can you declare a variable and method in jsp page?
5. what is the concept of mixing scriplets with html
6. What is expression in jsp explain with suitable example.
7. What is the syntax of el expression
8. What is implicit objects in EL expression
9. What are jsp literals?
10. What is the attributes of el expression.
11. What is the role of relational operator in jsp?
12. What is the difference b/w variable declared inside a declaration part and variable declared in scriplet part?

7
DIRECTIVES

7.1 INTRODUCTION OF DIRECTIVES

Directives are used to convey special information about the page container. Directives may be used to specify the scripting language for the page, to include the content of another page, or to indicate that the page uses a custom tag library. Directives do not directly produce an output that is visible to end-user when the page is requested. Instead, they generate side effects that change the way the jsp container processes the page.

JSP directives are as a message from a JSP page to the JSP container that control the processing of the entire page. They are used to set global values. All directives have scope to the entire JSP file. They are used to set page level instruction, insert data from files and specify custom tag libraries. JSP directives control how the JSP compiler generates the servlet. Actually, a directive is a way for you to give special instruction to the container at page translation time. The functionality of JSP directives is similar to the one provided by the C preprocessor.

A JSP directive affects the overall structure of the servlet class. It usually has the following form.

```
<%@ directive attribute="value" %>
```

However, you can also combine multiple attribute settings for a single directive as follows.

```
<%@ directive attribute1="value1" attribute2="value2" – attributeN="value N"
%>
```

7.2 ATTRIBUTE DIRECTIVE

Declares attributes of custom actions defined in tag files. Used in tag files only. The attribute directive allows the declaration of attributes for custom tags in tag files. This is analogous to the attribute element of a TLD file.

JSP syntax The JSP syntax for attribute Directive is given below.

```
<%@ attribute   name="attribute-name"
  [ required="true | false" ]
  [ fragment="true | false" ]
  [ rtexprvalue="true | false" ]
  [ type="java.lang.String | a non-primitive type"]
  [ description="text" ]
%>
```

or

```
<jsp:directive.attribute attributeDirectiveAttrList />
```

where attributeDirectiveAttrList is the same as the list for JSP syntax.

XML syntax The XML syntax for Attribute Directives is given below.

```
<jsp:directive.attribute attributeDirectiveAttrList />
```

where attributeDirectiveAttrList is the same as the list for JSP syntax.

Attributes The attributes of Attribute Directives are listed below.

name = 'attribute-name' The unique name of the attribute being declared. A translation error results if more than one attribute directive appears in the same translation unit with the same name. A translation error results if there is an attribute directive with a name attribute equal to the value of the name-given attribute of a variable directive or the dynamic-attributes attribute of a tag directive in this translation unit.

required='rue | false' Whether this attribute is required (true) or optional (false). Defaults to false if not specified.

fragment='true | false' Whether this attribute is a fragment to be evaluated by the tag handler (true) or a normal attribute to be evaluated by the container prior to being passed to the tag handler. If this attribute is true, then

i. you do not specify the rtexprvalue attribute. The container fixes the rtexprvalue attribute at true.

ii. you do not specify the type attribute. The container fixes the type attribute at javax.servlet.jsp.tagext.Jsp Fragment.

Defaults to false

rtexprvalue="true | false" Whether the attribute's value may be dynamically calculated at runtime by a scriptlet expression. Unlike the corresponding TLD element, this attribute defaults to true.

type="java.lang.String | a non-primitive type" The runtime type of the attribute's value. Defaults to java.lang.String if not specified. It is a translation error to specify a primitive type.

description="text" Description of the attribute. Defaults to no description.

***Example* 7.1** *Bookstore example*

The Bookstore example includes a custom tag, called ankDate, that is defined in a tag file, ankDate.tag. The tag file includes an attribute directive that declares the ankDate tag's attribute, shipping.

```
<%@ attribute name="shipping" required="true" %>
```

Here is an example of the ankDate tag used in the bookreceipt.jsp page.

```
<ankDate shipping="${param.shipping}" />
```

7.3 INCLUDE DIRECTIVE

The include directive inserts a file of text or code in JSP file at translation time, when the JSP is compiled. When the include Directive is used, the include process is static.

The included file can be an HTML file, a JSP page, a text file, XML document, or a code file written in the Java programming language. Be careful that the included file does not contain <html>, </html>, <body>, or </body> tags. Because the entire content of the included file is added to the including JSP page, these tags would conflict with the same tags in the including JSP page, causing an error.

Some of the behaviours of the include directive depends on the particular JSP container you are using, for example.

- The included file might be open and available to all requests, or it might have security restrictions.
- The JSP page might be recompiled if the included file changes.

JSP syntax The JSP syntax for Include Directive is given below.

```
<%@ include file="relativeURL" %>
```

or

```
<jsp:directive.include file="relativeURL" />
```

XML syntax The JSP syntax for include directive is given below.

```
<jsp:directive.include file="relativeURL" />
```

***Example* 7.2**

include.jsp: This example display result of date.jsp under include.jsp

```
<html>
<head> <title>An Include example</title> </head>
<body bgcolor="white">
<font color="blue">
The current date and time are:
<%@ include file="date.jsp" %>
</font>
</body>
</html>
```

date.jsp:

```
<%@ page import="java.util.*" %>
<%= (new java.util.Date() ).toLocaleString() %>
```

***Example* 7.3** file.html

This example display the static content of file.html under inc.jsp

```
<html>
```

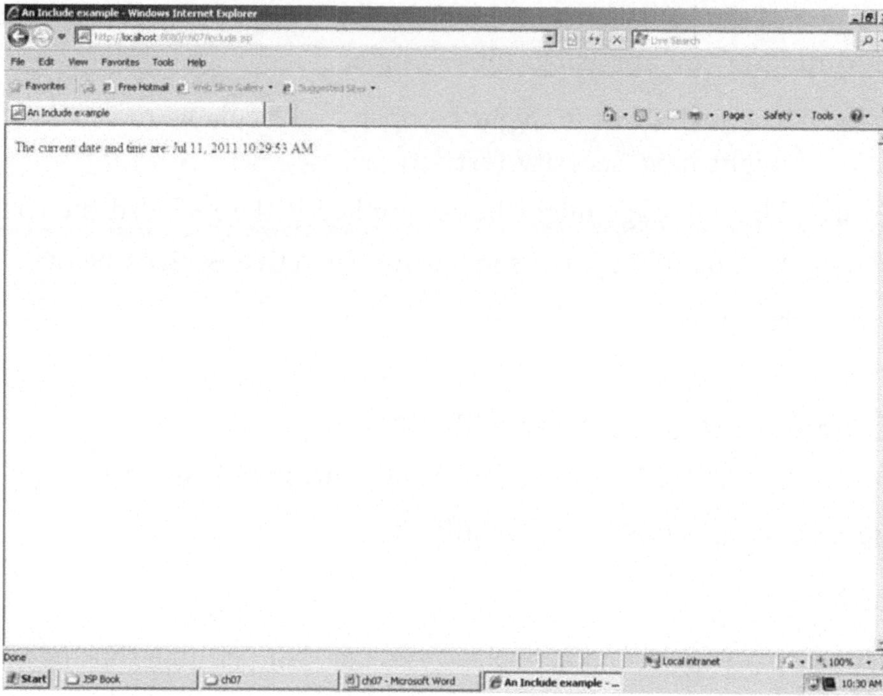

Figure 7.1 Output of include.jsp

```
<body>
<p>
this is the example for include directive</td> </tr>
</body>
</html>
```

inc.jsp

```
<html>
<head> <title>An Include example</title> </head>
<body bgcolor="white">
<font color="blue">
I think you got idea how this works
<%@ include file="file.html" %>
</font>
</body>
</html>
```

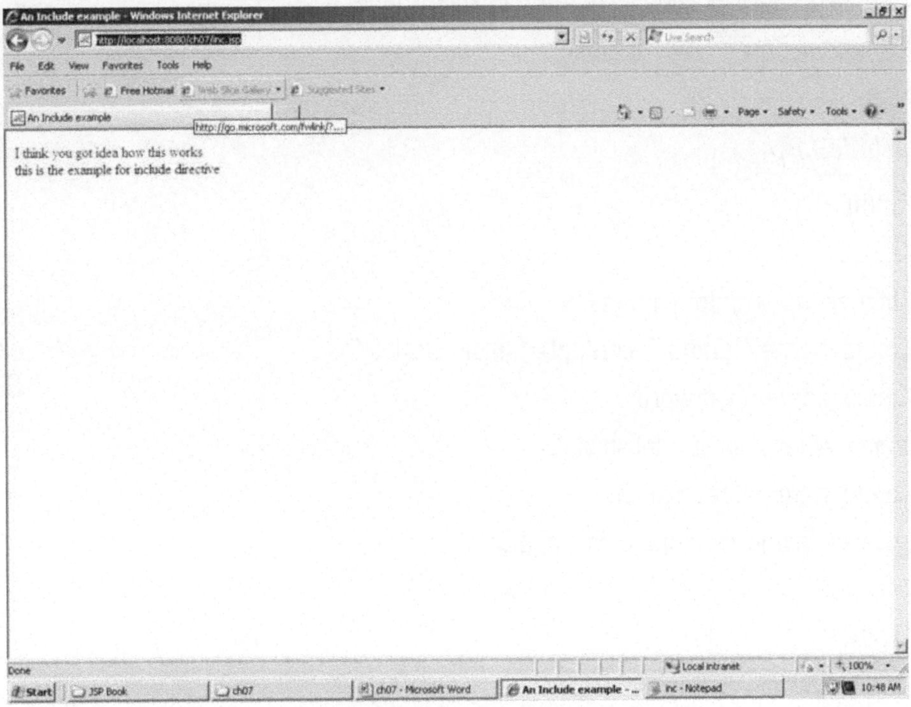

Figure 7.2 Output of inc.jsp

***Example* 7.4** Include2.jsp

```
<%@ page import="java.util.date" %>
<%-- the following become fields in each servlet that
results from a jsp page that includes this file. --%>
<%!
private int accesscount = 0;
private date accessdate = new date();
private string accesshost = "< I >no previous access</I >";
%>
<p>
<HR>
this page has been accessed <%= ++accessCount %>
times since server reboot. it was last accessed from
<%= accesshost %> at <%= accessDate %>.
```

```
<% accesshost = request.getremoteHost(); %>
<% accessdate = new date();
%>
```

Include3.jsp

```
<html>
<head>
<title>some random page</title>
<meta name="author" content="ANKUR SAXENA">
<meta name="keywords"
content="java,jsp,servlet,struts">
<meta name="description"
content="some random web page.">
</head>
<body>
<table border=5 align="center">
<tr><th class="title">
some random page</table>
<p>
information about our products and services.
<p>
java
<p>
java ,jsp,servlet
<%@ include file="include2.jsp" %>
</body>
</html>
```

***Example* 7.5** include4.jsp

This example display the content of header.jsp and footer.jsp

```
<%@page language="java" session="false" %>
<html>
```

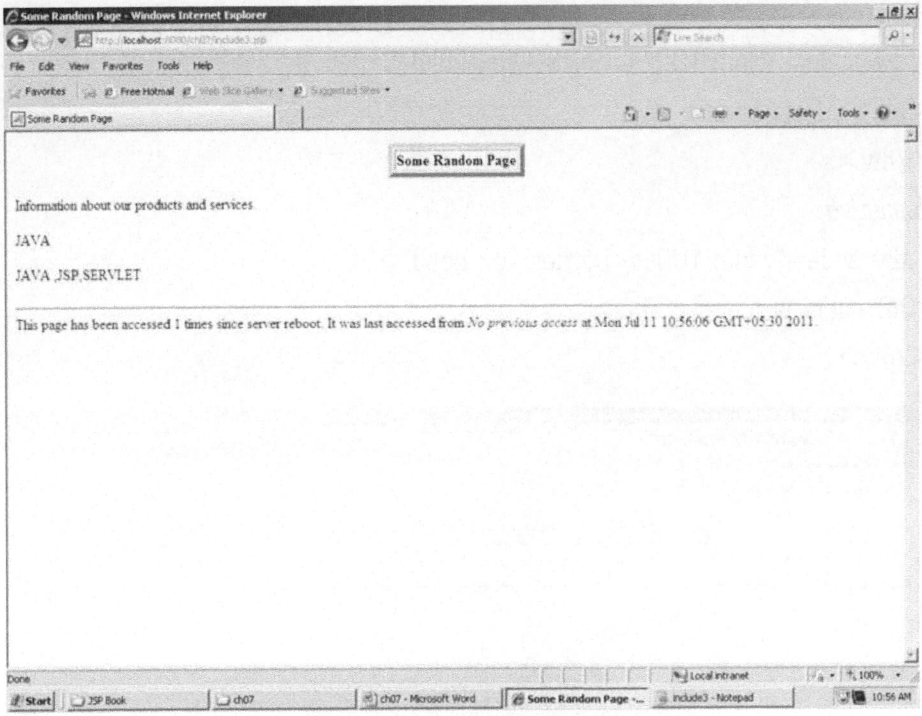

Figure 7.3 Output of include3.jsp

```
<head>
<title>the include directive with header and footer </title>
</head>
<body>
<div>
<%@include file="header.jsp"%>
<div style="padding:15% 33% 15% 33%; margin: 3px 0 3px 0; border:1px
    solid;">
<h2>hello ,ankur saxena </h2>
</div>
<%@include file="footer.jsp"%>
</div>
</body>
</html>
```

header.jsp

```
<div style="width:100%; border:1px solid;">
<h2> this is header </h2>
</div>
```

footer.jsp

```
<div style="width:100%; border:1px solid">
<h2> this is footer </h2>
</div>
```

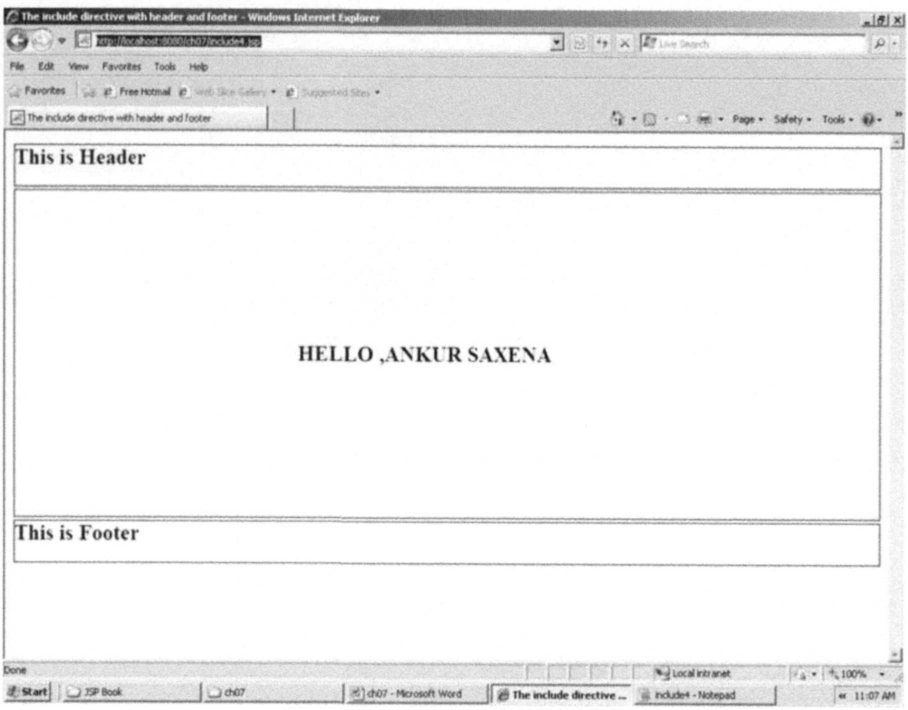

Figure 7.4 Output of include4.jsp

7.4 PAGE DIRECTIVE

The page directive applies to an entire JSP page and any of its static include files, which together are called a translation unit. A static include file is a file whose content becomes part of the calling JSP page. The page directive does not apply to any dynamic resources.

You can use the page directive more than once in a translation unit, but you can only use each attribute, except import, once. Because the import attribute is similar to the import statement in the Java programming language, you can use a page directive with import more than once in a JSP page or translation unit, no matter where you position the page directive in a JSP page or included files, it applies to the entire translation unit. However, it is often good programming style to place it at the top of the JSP page.

If you need to include a long list of packages or classes in more than one JSP page, you can create a separate JSP page with a page directive that contains the import list and include that file in the main JSP page.

JSP syntax The JSP syntax for page directive is given below.

```
<%@ page
  [ language="java" ]
  [ extends="package.class" ]
  [ import="{package.class | package.*}, ... " ]
  [ session="true|false" ]
  [ buffer="none|8kb|sizekb" ]
  [ autoFlush="true|false" ]
  [ isThreadSafe="true|false" ]
  [ info="text" ]
  [ errorPage="relativeURL" ]
  [ contentType="mimeType [ ; charset=characterSet ]" |
    "text/html ; charset=ISO-8859-1" ]
  [ isErrorPage="true|false" ]
  [ pageEncoding="characterSet | ISO-8859-1" ]
  [ isELIgnored="true|false"]
%>
```

or

```
<jsp:directive.page pageDirectiveAttrList />
```

where pageDirectiveAttrList is the same as the list in the JSP syntax.

XML syntax The XML Syntax for page directive is given below.

<jsp:directive.page pageDirectiveAttrList />

where pageDirectiveAttrList is the same as the list in the JSP syntax.

Attributes The attributes of page directive are listed below.

language

Syntax

```
<%@ page language = "lang" %>
<%@ page language = "java" %>
```

Language denotes, the scripting language used in scriptlets, declarations, and expressions in the JSP page.

extends This is used to signify name of the super class of the Java class used by the JSP engine.

Syntax

```
<%@ page extends = "package.class"%>
```

The fully qualified name of the super class of the Java class this JSP page will be compiled to use this attribute cautiously, as it can limit the JSP container's ability to provide a specialized super class that improves the quality of the compiled class.

import The import attribute is used to import all the classes in a java package into the current JSP page, because of this facility the JSP page can use other java classes.

Syntax

```
<%@ page import = "java.util.*" %>
```

If programmer wants to import more than one packages or classes, they can import the separate Java packages with commas:

```
<%@ page import="{package.class | package.*}, ... " %>
```

The following packages are implicitly imported, so you do not need to specify them with the import attribute.

java.lang.*

javax.servlet.*

javax.servlet.jsp.*

javax.servlet.http.*

You must place the import attribute before the element that calls the imported class.

Session The session attribute, when set to true, sets the page to make use of sessions. By default, the session attribute value is set to true. Therefore, all JSP pages have session data available.

If the user sets the session attribute to false, it should be declared false in attribute value.

Syntax

```
<%@ page session="true|false" %>
```

Whether the client must join an HTTP session in order to use the JSP page If the value is true, the session object refers to the current or new session.

If the value is false, you cannot use the session object or a <jsp:useBean> element with scope=session in the JSP page. Either of these usages would cause a translation-time error.

The default value is true.

Buffer The buffer size in kilobytes is, used by the out object to handle output sent from the compiled JSP page to the client web browser. The default value is 8kb. If you specify a buffer size, the output is buffered with at least the size you specified.

Syntax `<%@ page buffer=sizekb %>` Gives the size of the page buffer in kb or none for no buffer. If buffer is none, all output is immediately flushed. Default buffer size is 8kb.

auto flush Use to flush the page buffer when it fills. Default value is true. If value of autoFlush is false, the JSP engine will throw an exception if the buffer overflows.

Syntax

```
<%@ page autoFlush="true" %>
```

Whether the buffered output should be flushed automatically when the buffer is full If set to true (the default value), the buffer will be flushed. If set to false, an exception will be raised when the buffer overflows. You cannot set autoFlush to false when buffer is set to none.

is thread safe This tells the JSP that multiple pages can execute in parallel. Defaults to true.

Syntax

```
<%@ page isThreadSafe="true" %>
```

Whether thread safety is implemented in the JSP page The default value is true, which means that the JSP container can send multiple, concurrent client requests to the JSP page. You must write code in the JSP page to synchronize the multiple client threads. If you use false, the JSP container sends client requests one at a time to the JSP page.

- A text string that is incorporated verbatim into the compiled JSP page. You can later retrieve the string with the Servlet. getServletInfo() method.

  info="text"

- This defines a page of the given path to display if an error occurs in the JSP page.

Syntax

```
<%@ page errorPage="path" %>
```

A path name is a JSP page that this JSP page sends exceptions to. If the pathname begins with a /, the path is relative to the JSP application's document root directory and is resolved by the web server. If not, the path name is relative to the current JSP page.

iserror page This gives permission an error page access to the exception implicit variable. Its default value is false.

Syntax

```
<%@ page isErrorPage="true" %>
```

Whether the JSP page displays an error page If set to true, you can use the exception object in the JSP page. If set to false (the default value), you cannot use the exception object in the JSP page.

ContentType Sets the content type and character encoding of the page.

Syntax

```
<%@ page contentType="description" %>
```

The MIME type and character encoding the JSP page uses for the response. You can use any MIME type or character set that are valid for the JSP container. The default MIME type is text/html, and the default character set is ISO-8859-1.

pageEncoding="characterSet | ISO-8859-1" The character encoding the JSP page uses the response. The default character set is ISO-8859-1.

isElIgnored="true | false"

Example 7.6 page1.jsp

This example display the result of language attribute of page directive.

```
<%@page language="java" %>
<html>
<head> <title>Hello World JSP Page.</title> </head>
<body>
    <font size="10">
    <%
        String name="Ankur Saxena";
        out.println("Hello " + name + "!");
        %>
        </font>
    </body>
</html>
```

Example 7.7 page4.jsp

This example display the result of import attribute of page directive.

```
<%@page import="ankur.ex" %>
<html>
```

```
<head> <title>Example of Extends Attribute of page Directive in JSP</title> </head>
<body>
  <font size="20" color="red">
```

Figure 7.5 Output of page1.jsp

```
  <%
   ex ex1 = new ex();
   out.print(ex1.show());
  %>
  </font>
 </body>
</html>
```

Here is the code of ex.java file:

```
package ankur;
public class ex{
```

```
    public String show(){
    return "ANKUR SAXENA";
    }
  }
```

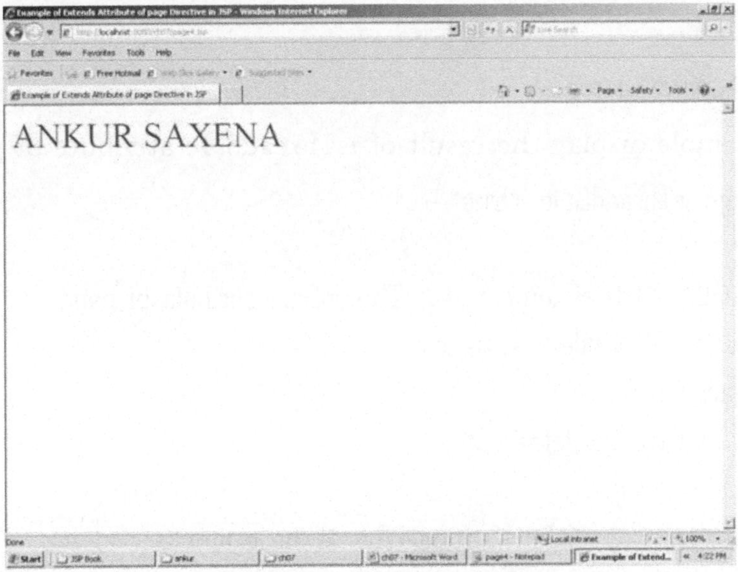

Figure 7.6 Output of page4.jsp

This code show all the attribute of page directive.

```
<%@page language="java" import="java.util.List, java.util.Date" Session="true"
    buffer="24kb" autoFlush="true" Info="JSP page directive
    errorPage="error.jsp" isErrorPage="false" isThreadSafe="true"
    contentType="text/html" pageEncoding="ISO-8859-1" />
```

***Example* 7.8**

This example display the result of isErrorPage attribute of page directive.

```
<%@ page  isErrorPage="true" %>
<html>
  <body>
    <!-- This displays the fully-qualified name of the exception
         and its message-->
    <%= exception.toString() %>
```

```
    <br>
    <!-- This displays the exception's descriptive message -->
    <%= exception.getMessage() %>
  </body>
</html>
```

***Example* 7.9** page5.jsp

This example display the result of isThreadSafe attribute of page directive.

```
<%@page isThreadSafe="true" %>
<html>
    <head> <title>Example of is ThreadSafe attribute of page
 directive in JSP</title> </head>
    <body>
        <table border="1">
            <tr>
                <td> <strong>This is the example
of isThreadSafe attribute of page directive.</td>
            </tr>
        </table>
    </body>
</html>
```

This code show the XML option of errorpage attributes.

```
<%@ page import="java.util.*, java.lang.*" %>
<%@ page buffer="5kb" autoFlush="false" %>
<jsp:directive.page errorPage="error.jsp" />
```

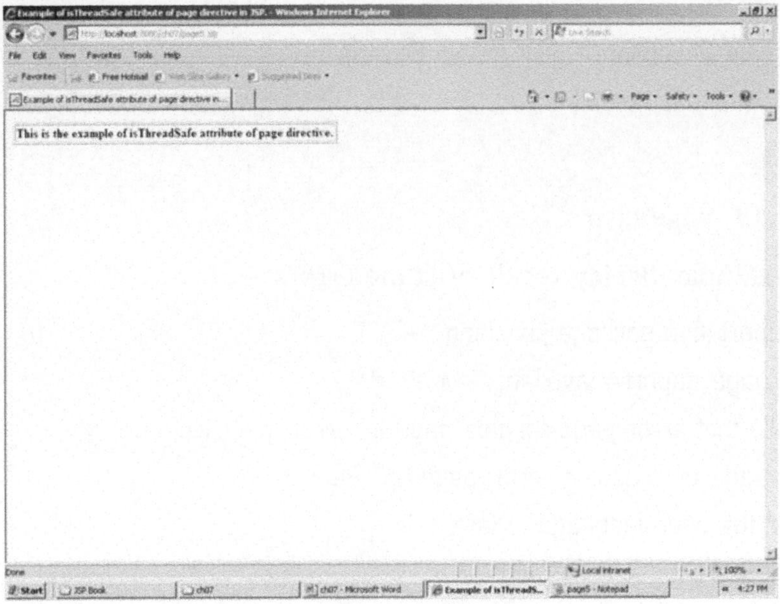

Figure 7.7 Output of page5.jsp

***Example* 7.10** *an2.JSP*

This example display the result of isErrorpage and error page attribute of page directive.

```
<html>
<body>
<%@page isErrorPage="true" %>
Error <%= exception.getMessage() %>
Has been reported
<html>
<body>
page7.jsp
<%@ Page errorPage="an2.jsp" %>
<%
if(true)
```

```
{
throw new Exception("Hello Ankur Saxena");
}
%>
```

Example 7.11 page9.jsp

This example display result of some errors.

```
<!-- import java package java.lang.* -->
<%@ page import="java.lang.*" %>
<!-- tells that error page for this page is 'error_page.jsp' -->
<%@ page errorpage="error_page.jsp" %>
<!-- set the page language to java -->
<%@ page language = "java" %>
<html>
   <head>
        <title>jsp directive </title>
   </head>
          <h2>some jsp page directives </h2>
   <body>
     <%
          out.println("in this example we have import java package<br>");
          out.println("and define a error page<br>");
          out.println("and define the langauage for this page <br>");
     %>
   </body>
</html>
```

error_page.jsp

```
<!-- tells that this jsp page is error page -->
<%@ page iserrorpage = "true"%>
<!-- import java package 'java.io' -->
```

```
<%@ page import = "java.io.*" %>
<HTML>
   <HEAD>
      <TITLE> custom error page </TITLE>
   </head>
<body>
<h2>your application has generated an error</h2>
<h3>please check for the error given below</h3>
<b>exception:</b> <br>
<font color="red"> <%= exception.tostring() %> </font>
</body>
</html>
```

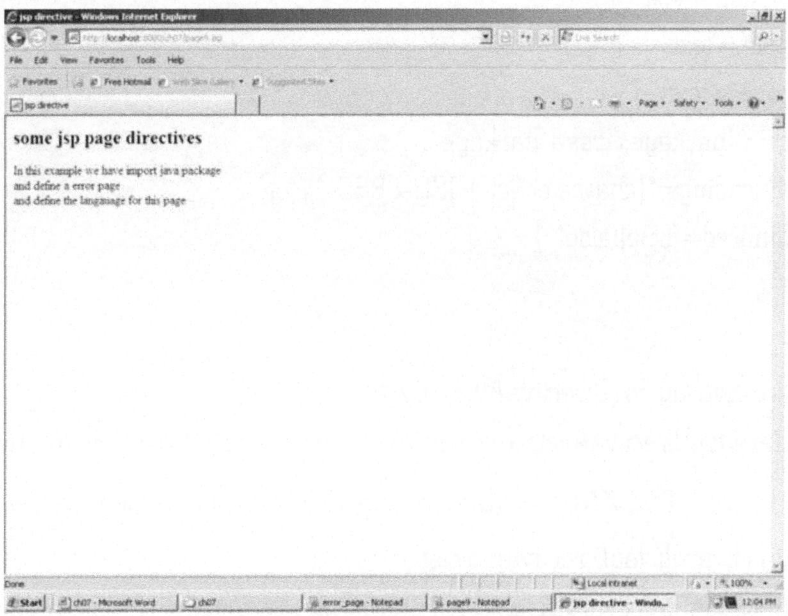

Figure 7.8 Output of page9.jsp

7.5 TAG DIRECTIVE

The tag directive is similar to the page directive in a JSP page, but applies to tag files instead of JSP pages. As with the page directive, a translation unit can contain more than one instance of the tag directive. All the attributes apply to the complete translation unit. However, there can be only one occurrence of any attribute or value defined by this directive in a given translation unit. With the exception of the import attribute, multiple attribute or value (re)definitions result in a translation error.

JSP syntax The JSP syntax of Tag directive is given below.

```
<%@ tag
[ display-name="name of the tag file | display-name" ]
[ body-content="scriptless|tagdependent|empty" ]
[ dynamic-attributes="page-scoped attribute" ]
[ small-icon="relativeURL" ]
[ large-icon="relativeURL" ]
[ description="text" ]
[ example="text" ]
[ language="java" ]
[ import="{package.class | package.*} , ... " ]
[ pageEncoding="{characterSet | ISO-8859-1}" ]
[ isELIgnored="true|false" ]
%>
```

or

```
<jsp:directive.tag tagDirectiveAttrList />
```

where tagDirectiveAttrList is the same as the attribute list in the JSP syntax.

XML syntax The XML syntax of Tag directive is shown below.

```
<jsp:directive.tag tagDirectiveAttrList />
```

where tagDirectiveAttrList is the same as the attribute list in the JSP syntax.

Attributes The Attributes of Tag directive are listed below.

display-name-"name of the tag file | display-name" A short name that is intended to be displayed by tools. Defaults to the name of the tag file, without the .tag extension.

body-content-'scriptless|tagdependent|empty' Provides information on the content of the body of this tag. Can be either empty, tag-dependent, or scriptless. A translation error results if JSP or any other value is used. Defaults to scriptless.

dynamic-attributes-"page-scoped attribute" Indicates whether this tag supports additional attributes with dynamic names. If present, the generated tag handler must implement the javax.servlet.jsp.tagext. Dynamic Attributes interface, and the container must treat the tag as if its corresponding TLD entry contained<dynamic-attributes>true</dynamic-attributes>. The implementation must not reject any attribute names. A Map is an object that maps keys to values. A Map cannot contain duplicate keys: each key can map to at most one value. The value identifies a page-scoped attribute in which to place a Map containing the names and values of the dynamic attributes passed during invocation of the tag. The Map must contain each dynamic attribute name as the key and the dynamic attribute value as the corresponding value. Only dynamic attributes with no URI are to be present in the Map, all other dynamic attributes are ignored. A translation error results if the value of the dynamic-attributes of a tag directive is equal to the value of a name-given of a variable directive or the value of a name attribute of an attribute directive.

small-icon-"relativeURL" Either a context-relative path, or a path relative to the tag source file of an image file containing a small icon that can be used by tools. Defaults to no small icon.

large-icon-"relativeURL" Either a context-relative path, or a path relative to the tag source file of an image file containing a large icon that can be used by tools. Defaults to no large icon.

description-"text" Defines an arbitrary string that describes this tag. Defaults to no description.

example-"text" Defines an arbitrary string that presents an informal description of an example of a use of this action. Defaults to no example.

language-"java" Carries the same syntax and semantics of the language attribute of the page directive.

import-"{package.class | package.}, ..."* Carries the same syntax and semantics of the import attribute of the page directive.

pageEncoding-"{characterSet | ISO-8859-1}" Carries the same syntax and semantics of the pageEncoding attribute in the page directive. However, there is no corresponding global configuration element in web.xml. The pageEncoding attribute cannot be used in tag files in XML syntax.

isELIgnored-"true|false" Carries the same syntax and semantics of the isELIgnored attribute of the page directive. However, there is no corresponding global configuration element in web.xml.

Example 7.12 Color map

This tag accepts an arbitrary number of attributes whose values are colors and outputs a bulleted list of the attributes colored according to the values

```
<colored:colored color1="red" color2="yellow" color3="blue"/>
```

The following code implements the preceding tag. An arbitrary number of attributes whose values are colors are stored in a map named by the dynamic-attributes attribute of the tag directive. The JSTL for each tag is used to iterate through the map and the attribute keys and colored attribute values are printed in a bulleted list.

```
<%@ tag dynamic-attributes="colorMap"%>
<ul >
<c:forEach var="color" begin="0" items="${colorMap}">
< li >${color.key} = <font color="${color.value}">${color.value}
</font> <li>
</c:forEach>
</ul>
```

This code show the attributes of taglib directive.

```
<%@ tag name="msg"
  display-name="Message"
  body-content="scriptless"
  dynamic-attributes="user"
```

```
  small-icon="/WEB-INF/small-icon.jpg"
  large-icon="/WEB-INF/large-icon.jpg"
  description="Simple usage of a tag directive"
%>
```

7.6 TAGLIB DIRECTIVE

The taglib directive declares that your JSP page uses a set of custom tags, identifies the location of the library, and provides a means for identifying the custom tags in your JSP page.

You must use a taglib directive before you use the custom tag in a JSP page. You can use more than one taglib directive in a JSP page, but the prefix defined in each must be unique.

Taglib directive is used to define the tag library which is a collection of tags. A Tag library can be a standard(pre-defined) or a custom(user-defined) one. The "prefix" attribute is a must for custom as well as standard tags in jsp.

JSP syntax The JSP syntax of Taglib directive is given below.

```
<%@ taglib {uri="URI" | tagdir="/WEB-INF/tags[/subdir]+"}
prefix="tagPrefix" %>
```

XML syntax The XML syntax of Taglib directive is given below.

```
<anyxmlelement xmlns:prefix="{uri | urn:jsptld:path |
   urn:jsptagdir:/WEB-INF/tags[/subdir]+}" >
```

Attributes The attributes of Taglib directive is listed below.

uri-"URI" The Uniform Resource Identifier (URI) that uniquely locates the TLD that describes the set of custom tags associated with the named tag prefix. A URI can be any of the following.

- A Uniform Resource Locator (URL), as defined in RFC 2396, available at http://www.hut.fi/u/jkorpela/rfc/2396/full.html
- A Uniform Resource Name (URN), as defined in RFC 2396
- An absolute or relative pathname

If the URI is a URL or URN, then the TLD is located by consulting the mapping indicated in web.xml extended using the implicit maps in the packaged tag libraries. If URI is pathname, it is interpreted relative to the root of the web application and should resolve to a TLD file directly, or to a JAR file that has a TLD file at location META-INF/taglib.tld.

tagdir-" /WEB-INF/tags[/subdir]+" Indicates this prefix is to be used to identify tag extensions installed in the /WEB-INF/tags/ directory or a subdirectory. An implicit tag library descriptor is used. A translation error must occur under any of these conditions.

- The value does not start with /WEB-INF/tags/.
- The value does not point to a directory that exists.
- If this attribute is used in conjunction with the uri attribute.

prefix "tagPrefix" The prefix that precedes the custom tag name, for example, public in <public:loop>. Empty prefixes are illegal. If you are developing or using custom tags, you cannot use the tag prefixes jsp, jspx, java, javax, servlet, sun, and sunw, as they are reserved by Sun Microsystems.

```
 <%@ taglib uri="tlds/taglib.tld" prefix="mytag" %>
<%@ taglib uri="http://www.jspcentral.com/tags" prefix="public" %>
<public:loop>
  ...
</public:loop>
```

***Example* 7.13**

This example display the multiplication of two number using taglib directive.

```
<%@ taglib uri="http://java.sun.com/jsp/jstl/core" prefix="c" %>
<p>Simple Example for C:out</p>
 Multiply 5 and 2::<c:out value="${5*2}" />
```

Result

```
Simple Example for C:out
Multiply 5 and 2::10
```

Example 7.14

For example, suppose the custlib tag library contains a tag called hello. If you wanted to use the hello tag with a prefix of mytag, your tag would be <mytag:hello> and it will be used in your JSP file as follows:

```
<%@ taglib uri="http://www.example.com/custlib" prefix="mytag" %>
<html>
<body>
<mytag:hello/>
</body>
</html>
```

7.7 VARIABLE DIRECTIVE

The variable directive is used to declare variables in tag files. The variable directive is analogous to the variable element in the Tag Library descriptor, and defines the details of a variable exposed by the tag handler to the calling page.

A custom tag (whether it's implemented using Java or a tag file) is able to declare that it returns variables to the calling page.

JSP syntax The JSP syntax of variable directive is given below.

```
<%@ variable { name-given="scripting variable" |
  (name-from-attribute="scripting variable"
  alias="locally-scoped attribute")}
  [ variable-class="java.lang.String" | name of the variable class" ]
  [ declare="true | false" ]
  [ scope="AT_BEGIN | AT_END | NESTED" ]
  [ description="text" ]
%>
```

or

```
<jsp:directive.variable variableDirectiveAttrList />
```

XML Syntax The XML syntax of variable directive is given below.

```
<jsp:directive.variable variableDirectiveAttrList />
```

Attributes The attributes of variable directive are listed below.

*name-given="scripting variable" | name-from-attribute="scripting variable"

Defines an EL variable to be used in the page invoking this tag. Either name-given or name-from-attribute must be specified. If name-given is specified, the value is the name of an attribute whose (translation-time) value at the start of the tag invocation will give the name of the variable.

Translation errors arise in the following circumstances.

- Specifying neither name-given nor name-from-attribute.
- If two variable directives have the same name-given.
- If the value of a name-given attribute of a variable directive is equal to the value of a name attribute of an attribute directive or the value of a dynamic-attributes attribute of a tag directive.

alias-"locally-scoped attribute" Defines a variable, local to the tag file, to hold the value of the EL variable. The container will synchronize this value with the variable whose name is given in name-from-attribute.

Required when name-from-attribute is specified. A translation error results if used without name-from-attribute.

A translation error results if the value of alias is the same as the value of a name attribute of an attribute directive or the name-given attribute of a variable directive.

variable-class-"java.lang.String" | name of the variable class" The name of the class of the variable. The default is java.lang.String.

declare-" true | false" Whether or not the variable is declared. Defaults to true.

scope-' AT_BEGIN | AT_END | NESTED' The scope of the variable. Defaults to NESTED.

description-" text"

Example 7.14

The following tag file declares a variable, x, with a scope of AT_END and sets its value to 3.

```
<%-- a.tag --%>
<%@ variable name-given=" x" scope=" AT_END" %>
<c:set var="x" value="3" />
```

The following JSP page references the variable with an EL expression.

```
<%-- b.jsp --%>
<%@ taglib prefix="tags" tagdir="/WEB-INF/tags" %>
${x}
<tags:a />
${x}
```

The first reference to the variable is undefined because the variable has not been declared yet. After the tag file has been invoked and the variable is set, the page references the variable again, which will result in an output of 3.

This code show how to place information inside bean.

The following tag can look up user information and place it in a bean.

```
<mytag:lookupUserInfo />
${userInfo.name}
```

In this case, the lookupUserInfo tag can declare that it returns a variable to the calling page with a name of userInfo.

Here is another example

```
<mytag:displayCustomers>
   ${customer.name}
</mytag:displayCustomers>
```

The scope of the variable affects how it is exposed to the calling page (AT_BEGIN means the variable is available from the start tag onwards, AT_END means the variable is available from the end tag onwards, and NESTED means the variable is only available in the body of the tag). In the first preceding example, you would probably use AT_END so that the variable is still available at the end of the tag. In the second example, you would probably use NESTED since the ${customer} variable does not need to be accessible anywhere but inside the displayCustomers tag.

When you want to emulate OUT parameters, use variables with scope AT_BEGIN or AT_END. For each AT_BEGIN or AT_END variable, a page-scoped attribute is made available in the JspContext of the tag file. The scoped attribute is not initialized. Synchronization is performed at the end of the tag for AT_BEGIN and AT_END and also before the invocation of a fragment for AT_BEGIN.

When you want to emulate nested parameters, use variables with scope AT_BEGIN or NESTED. For each AT_BEGIN or NESTED variable, a page-scoped attribute is made available in the JspContext of the tag file. The scoped attribute is not initialized. Synchronization is performed before each fragment invocation for AT_BEGIN and NESTED, and also after the end of the tag for AT_BEGIN.

The JSP specification recommends that to accomplish IN parameters, use attributes. To accomplish OUT or NESTED parameters, use variables.

REVIEW QUESTIONS

1. What is the role of directive in page container?
2. What is the difference between XML and JSP syntax of attribute directive?
3. What is include directive?
4. How do I include static files within a JSP page?
5. What are the various tags available?
6. What is the role of page directive in dynamic resources?
7. Write about the attribute of a page directive
8. Explain session in page directive.
9. Explain the concept of tag directive.
10. What is the difference between tag and taglib directive.
11. How to introduce variable directive in your JSP page.
12. How can I implement a thread-safe JSP page? What are the advantages and Disadvantages of using it?

8

STANDARD ACTIONS

8.1 INTRODUCTION

JSP Standard actions are pre-defined tags which perform some action based on information that is required at the exact time the JSP page is requested by a browser. An action can, for instance, access parameters sent with the request to do a database lookup. It can also dynamically generate HTML, such as a table filled with information retrieved from an external system. They can be used to include a file at the request time, to find or instantiate JavaBeans, to forward a request to a new page, to generate a browser-specific code, etc. The JSP standard actions affect the overall runtime behaviour of a JSP page and also the response sent back to the client.

8.2 jsp:attribute

The jsp:attribute element allows you to define the value of a tag attribute in the body of an XML element instead of the value of an XML attribute.

The jsp:attribute standard action has two uses.

- It allows the page author to define the value of an action attribute in the body of an XML element instead of the value of an XML attribute.
- It allows the page author to specify the attributes of an element dynamically generated by the jsp:element action, which can only appear as a subelement of a standard or custom action.

All JSP standard actions and custom actions can contain a jsp:attribute standard element as a substitute for any of its attributes. One use-case in which jsp:attribute is particularly helpful is where the value of an attribute is the result of a multi-line expression, which would not fit in the value of an attribute in the start tag of the action.

If an action contains any jsp:attribute elements and the action also has a body, it must use the jsp:body tag to represent the body. The body of jsp:attribute is restricted according to the type of attribute being specified.

- For simple attributes that accept an EL expression, the body can be any JSP content.
- For simple attributes that do not accept an EL expression, the body can contain only static text.
- For fragment attributes, the body must not contain any scripting elements.

JSP syntax The JSP syntax for jsp:attribute is given below.

<jsp:attributename="attributeName" [trim= "true | false"] />

Attributes The attribute of jsp:attribute is listed below.

name="attributeName" Identifies which tag attribute is being specified. When used with jsp:element, this attribute specifies the name of the attribute to be included in the generated element.

If the jsp:attribute tag is not being used with jsp:element and the action does not accept dynamic attributes, the name must match the name of an attribute for the action being invoked, as declared in the Tag Library Descriptor (TLD) for a custom action, or as specified for a standard action. If it does not match, a translation error results.

If the jsp:attribute element is being used with a jsp:element tag, a translation error results if both an XML element attribute and a jsp:attribute element are used to specify the value for the same attribute.

The value of name can be a QName. If so, a translation error must occur if the prefix does not match that of the action it applies to, unless the action supports dynamic attributes or the action is jsp:element.

trim="true|false" The optional trim attribute determines whether or not whitespace appearing at the beginning and end of the element body should be discarded. By default, the leading and trailing whitespace (including spaces, carriage returns, line feeds, and tabs) is discarded. The whitespace is trimmed when the JSP page is translated. If a body contains a custom tag that produces leading or trailing whitespace, that whitespace is preserved regardless of the value of the trim attribute.

The following template page uses jsp:attribute, which uses the output of fmt:message to set the value of the value attribute of tt:parameter where fmt, tt are custom tag prefix).

```
<%@ taglib prefix="tt"
uri="WEB-INF/custom.tld"%>
<%@ taglib prefix="fmt"
uri="WEB-INF/custom.tld"%>
<tt:screenid="/jspbook">
<tt:parameter name="title" direct="true">
<jsp:attribute name="value" >
<fmt:message key="TitleBook"/>
</jsp:attribute>
</tt:parameter>
...
</tt:screen>
...
```

8.3 jsp:body

Explicitly defines an element body. This tag is required if the enclosing element contains any jsp:attribute tags. Normally, the body of a standard or custom action invocation is defined implicitly as the body of the XML element used to represent the invocation. The body of a standard or custom action can also be defined explicitly using the jsp:body standard action. This is required if one or more jsp:attribute elements appear in the body of the tag.

If one or more jsp:attribute elements appear in the body of a tag invocation but no jsp:body element appears or an empty jsp:body element appears, it is equivalent to the tag having an empty body.

It is also legal to use the jsp:body standard action to supply bodies to any standard actions that accepts a body (except for jsp:body, jsp:attribute,jsp:scriptlet,jspexpression,and jsp:declaration). The body standard action accepts no attributes.

JSP syntax The JSP syntax of jsp: body is given below:

```
<jsp:body>any elements or text</jsp:body>
```

8.4 jsp:element

The jsp:element action is used to dynamically define the value of the tag of an XML element. This action can be used in JSP pages, tag files and JSP documents.

A jsp:element action has one mandatory attribute, name, of type String. The value of the name attribute is used as that of the tag of the element generated. The jsp:element action can have a body. Two forms are valid, depending on whether the element is to have attributes or not.

In the first form, no attributes are present.

```
<jsp:elementname="name">
  optional body
</jsp:element>
```

In the second form, zero or more attributes are requested, using jsp:attributeand jsp:body, as appropriate.

```
<jsp:element name="name">
  [jsp:attribute]+
  [jsp:body]
</jsp:element>
```

The one valid, mandatory, attribute of jsp:element is its name. Unlike other standard actions, the value of the name attribute must be given as an XML-style attribute and cannot be specified using jsp:attribute. This is because jsp:attribute has a special meaning when used in the body of jsp:element.

JSP syntax The JSP syntax for jsp:element is given below.

```
(<jsp:element name="elementName"
 (/>|>(any elements or text</jsp:element>))
)
|(<jsp:element name="elementName">
      [<jsp:attribute name="attributeName"
        [trim="true|false" ]
     (/>|(any elements or text </jsp:attribute>))]+
     [<jsp:body> any elements or text </jsp:body>]
   </jsp:element>
 )
```

Attributes The attributes of jsp:element is listed below.

name="elementName" The value of name is that of the element generated. The name can be a QName. JSP places no constraints on this value, it is accepted as it is. A request-time attribute value may be used for this attribute.

Example 8.1

The <jsp:element>, <jsp:attribute> and <jsp:body> actions are used to define XML elements dynamically. The word dynamically is important, because it means that the XML elements can be generated at request time rather than statically at compile time.

Following is a simple example to define XML elements dynamically.

```
<%@page language="java" contentType="text/html"%>
<html xmlns="http://www.w3c.org/1999/xhtml"
  xmlns:jsp="http://java.sun.com/JSP/Page">
<head> <title>Generate XML Element</title> </head>
<body>
<jsp:element name="xmlElement">
<jsp:attribute name="xmlElementAttr">
  Value for the attribute
</jsp:attribute>
```

```
<jsp:body>
  Body for XML element
</jsp:body>
</jsp:element>
</body>
</html>
```

This would produce following HTML code at run time.

```
<html xmlns="http://www.w3c.org/1999/xhtml" xmlns:jsp="http://java.sun.com/
    JSP/Page">
 <head> <title> Generate XML Element </title> </head>
<body>
<xmlElement xmlElementAttr="Value for the attribute">
  Body for XML element
</xmlElement>
</body>
</html>
```

8.5 jsp:doBody

Evaluates the body of the tag used by the calling page to invoke this tag file.

The jsp:doBody standard action can only be used in tag files. It invokes the body of the tag, sending the output of the result to the JspWriter, or to a scoped attribute that can be examined and manipulated.

The jsp:doBody standard action behaves exactly like jsp:invoke, except that it operates on the body of the tag instead of a specific fragment passed as an attribute. Because it always operates on the body of the tag, there is no name attribute for this standard action.

Fragments are provided access to variables the same way for jsp:doBody as they are for jsp:invoke. If no body was passed to the tag, jsp:doBody will behave as though a body was passed in that produces no output.

The body of a tag is passed to the simple tag handler as a JspFragment object. A translation error shall result if the jsp:doBody action contains a non-empty body.

The result of evaluating the tag body is sent to the response or is stored in an EL variable for later manipulation. To store the result of evaluating a fragment to an EL variable, you specify the var or varReader attribute. If var is specified, the container stores the result in an EL variable of type String with the name specified by var. If varReader is specified, the container stores the result in an EL variable of type java.io.Reader, with the name specified by varReader. The Reader object can then be passed to a custom tag for further processing. A translation error occurs if both var and varReader are specified. An optional scope attribute indicates the scope of the resulting variable. The possible values are page (default), request, session, or application. A translation error occurs if you use this attribute without specifying the var or varReader attribute.

JSP syntax The JSP syntax of jsp: doBody is given below.

```
<jsp:doBody({var="scopedAttributeName"|
     varReader="scopedAttributeName"}
 [scope="page|request|session|application"]/>)|/>.
```

Attributes The attributes of Jsp:do Body are listed below.

var="scopedAttributeName" The name of a scoped attribute to store the result of the body invocation in, as a java.lang.String object. A translation error must occur if both var and varReader are specified. If neither var nor varReader are specified, the result of the body goes directly to the JspWriter, as described above.

varReader="scopedAttributeName" The name of a scoped attribute to store the result of the body invocation in, as a java.io.Reader object. A translation error must occur if both var and varReader are specified. If neither var nor varReader is specified, the result of the body invocation goes directly to the JspWriter, as described above.

scope="page|request|session|application" The scope in which to store the resulting variable. A translation error results if the value is not one of page, request, session, or application. A translation error results if this attribute appears without specifying either the var or varReader attribute as well. Note that a value of session should be used with caution since not all calling pages may be participating in a session. A container must throw an IllegalStateException at runtime if scope is session and the calling page does not participate in a session. Defaults to page.

This example tag file encapsulates the double custom tag.

```
<%--double.tag --%>
<jsp:doBody/>
<jsp:doBody/>
```

It includes two jsp:doBody tags, which will each invoke the body of the custom double tag in any JSP that includes the tag.

This example JSP page uses the custom double tag to output the text included in the body of the tag two times.

```
<%-- a.jsp --%>
<%@ taglibprefix="tags"tagdir="/WEB-INF/tags"%>
<tags:double>
 Hello Ankur Saxena<br>
</tags:double>
```

The result will be the output.

Hello Ankur Saxena

Hello Ankur Saxena

8.6 jsp:forward

The jsp:forward element forwards the request object containing the client request information from one JSP page to another resource. The target resource can be an HTML file, another JSP page, or a servlet, as long as it is in the same application context as the forwarding JSP page. The lines in the source JSP page after the jsp:forward element are not processed.

You can pass parameter names and values to the target resource by using a jsp:param clause. An example of this would be passing the parameter name username (with name="username") and the value scott (with value="scott") to a servlet as part of the request. If you use jsp:param, the target resource should be a dynamic resource that can handle the parameters.

Be careful when using jsp:forward with unbuffered output. If you have used the page directive with buffer="none" to specify that the output of your JSP page it should not be buffered, and if the JSP page has any data in the out object, using jsp:forward will cause an IllegalStateException.

JSP syntax The JSP syntax of jsp: forward is given below.

```
<jsp:forwardpage="{relativeURL|'${'Expression'}'|
<%=expression %>}"{> |> [ < jsp:param name="parameterName"
value="{parameterValue|'${'Expression'}'| <%= expression %> }"}
/> ]+ </jsp:forward> }
```

XML syntax The XML syntax of jsp: forward is given below.

```
<jsp:forwardpage="{relativeURL|'${'Expression'}'|
%=expression %}"{/> |> [ < jsp:param name="parameterName"
  value="{parameterValue|'${'Expression'}'|%= expression %}"}
/ > ]+ < /jsp:forward> }
```

Attributes The attributes of jsp: forward are listed below.

page="{relativeURL|<%=expression%>}" A String or an expression representing the relative URL of the component to which you are forwarding the request. The component can be another JSP page, a servlet, or any other object that can respond to a request.

The relative URL looks like a path, it cannot contain a protocol name, port number, or domain name. The URL can be absolute or relative to the current JSP page. If it is absolute (beginning with a /), the path is resolved by your web or application server.

<jsp:param name="parameterName"value="{parameterValue|<%= expression%>}" />+ The above code sends one or more name/value pairs as parameters to a dynamic resource. The target resource should be dynamic, that is, a JSP page, servlet, or other resource that can process the data that is sent to it as parameters.

You can use more than one jsp:param clause if you need to send more than one parameter to the target resource. The name attribute specifies the parameter name and takes a case-sensitive literal string as a value. The value attribute specifies the parameter value and takes either a case-sensitive literal string or an expression that is evaluated at request time.

The general syntax of this action is as follows.

```
<jsp:forwardpage="Relative URL"/>
```

8.6.1 jsp:forward vs jsp:include

jsp: forward behaves in a similar manner to the action jsp: include. The only difference is that the control is not returned to the calling page, so the code that follows the action is ignored.

Other differences are that the object request is passed to the new page. So any buffered messages are deleted. In case unbuffered output is thrown because the HTTP protocol is not able to handle the request if the transfer of part of the data (specifically, the HTTP headers) have already been sent to the client.

In a sense, this action has the same effect as the method sendRedirect Item response. Both methods redirect the user to a new resource, but it is important to understand how they work technically.

The method sendRedirect send code to the browser that states to call a new resource and this means that there is a continuous exchange of replies between clients and servers. Moreover, in the context redirect you miss the request object, and then any values to pass to the new resources should be managed differently.

While the action jsp: forward requests are redirected without going to the client and this results in a significant increase in performance.

Example 8.2

The above jsp page forwards the request to welcome.jsp page by using <jsp:forward page="welcome.jsp"/>

for.jsp

```
<%@ pagelanguage ="java" %>
<html>
<head> <title>jsp forwarding a request </title> </head>
<jsp:forward page="welcome.jsp"/>
</html>
```

Here is the code of welcome.jsp

```
<%@page language="java" %>
<html>
```

```
<head> <title>jsp forwarding a request</title> </head>
<body>
<h3> <b>Welcome!</b> </h3>
This is a JSP forward request
Request is forward here.
</body>
</html>
```

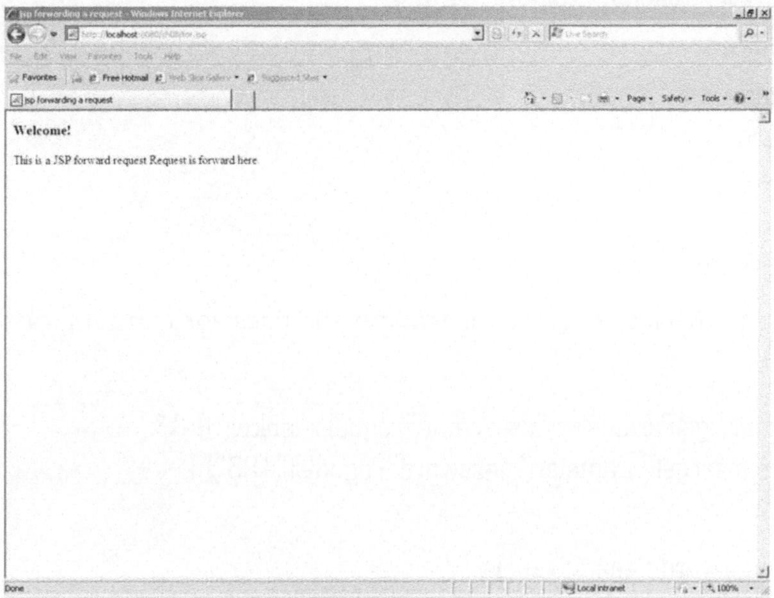

Figure 8.1

This screen (Figure 8.1) show the result of for.jsp that forward the request to welcome .jsp

***Example* 8.3** an4.html

This code create a username, password and post to an4.jsp

```
<html>
<head>
<title>
Action Test Page
```

```
</title>
</head> <center>
<form method="post" action="an4.jsp">
User Name<input type="text" name="username">
<br>
Password<input type="password" name="password">
<br>
<input type="submit" value="log in">
</form>
</center>
</body>
</html>
```

***Example* 8.4** an4.jsp

This code check the correct username and password and forward to an5.jsp

```
<%
  if((request.getParameter("username").equals("ankur")) &&
      (request.getParameter("password").equals("123")))
  {%>
    <jsp:forward page= "an5.jsp"/>
    <%}else {%>
<%@ include file ="an4.html" %>
<br> <center>
Plz check your username and password.
<br>
Try Again.
</center>
<%} %>
```

Figure 8.2

This screen (Figure 8.2) display the username and password. If username and password are correct then show next screen otherwise display try again.

***Example* 8.5** an5.jsp

This code display login information

```
<html>
<body>
<%! String s;%>
Login Successfully.<br>
You are valid user.<br>
welcome,<%=request.getParameter("username") %>
<br>
<% s=request.getParameter("password"); %>
Your password is:-
<%
```

```
  out.println(s);
%>
</body>
</html>
```

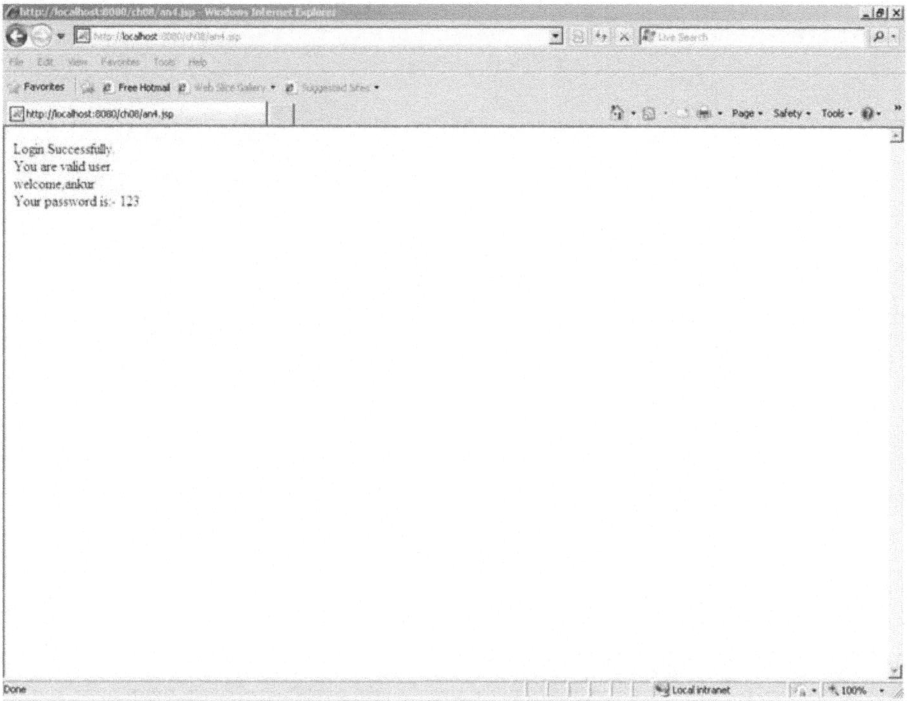

Figure 8.3 Output of an5.jsp

This screen (Figure 8.3) display the correct information of username and password

The following code display how to pass the value of username using action.

```
<jsp:forward page="/servlet/login"/>
<jsp:forward page="/servlet/login">
  <jsp:param name="username" value="jsmith"/>
</jsp:forward>
```

***Example* 8.6**

This sample code shows the use of tag. This checks the percentage of free memory and based on that opens new page using this tag.

memory.jsp

```
<html>
<%
double freeMemory = Runtime.getRuntime().freeMemory();
double totalMemory = Runtime.getRuntime().totalMemory();
double percent = freeMemory/totalMemory;
if(percent<0.5){
%>
<jsp:forward page="one.jsp"/>
<%}else{%>
<jsp:forward page="two.jsp"/>
<%}%>
</html>
```

one.jsp

```
<html>
<body>
VM Memory usage less than 50 percent
</body>
</html>
```

two.jsp

```
<html>
<body>
VM Memory usage greater than 50 percent
</body>
</html>
```

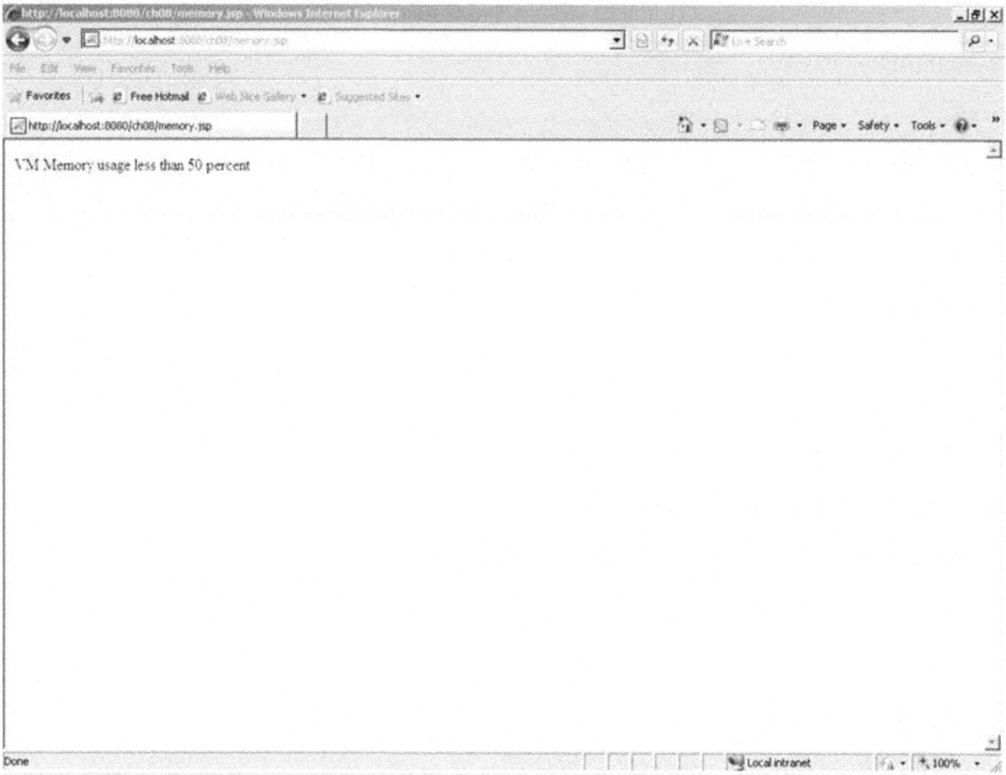

Figure 8.4 Output of memory.jsp

This screen (Figure 8.4) display the result of memory.jsp which forward the request of two jsp like one.jsp and two.jsp

8.7 jsp:usebean

The <jsp:useBean> is a standard action element used to locate or instantiate JavaBeans component. Firstly, <jsp: useBean> tries to locate an instance of the Bean class, if found its fine, if not then it will instantiate it from a class. The name of the bean is same as we have given in id attribute of <jsp:useBean>. If the object reference does not exist with the name we have specified then it will create an instance and find the scope of the variable, class attributes defines the bean class and type attribute defines the parent class or interface of the Bean class.

JSP syntax The JSP syntax for jsp:useBean is given below.

```
<jsp:useBean id= "nameOfInstance"scope="page|request|session|application"class=
    "package.class" type="package.class > </jsp:useBean >
```

***Example* 8.7**

```
<jsp:useBean id="cart" scope="session" class="session.Carts"/>
<jsp:setProperty name="cart" property="*"/>
<jsp:useBean id="checking" scope="session" class="bank.Checking">
  <jsp:setProperty name="checking" property="balance" value="0.0"/>
</jsp:useBean>
```

8.8 jsp:setProperty

This is used to set the value of one or more properties of bean using setter method. The value of the name must match with the Id.

JSP syntax The JSP syntax of jsp: setproperty is given below.

```
<jsp:setPropertyname="beanInstanceName"
{
property="*"|
property="propertyName"[param="parameterName"]|
property="propertyName"value="{stringLiteral|
'${'Expression'}'|<%= expression %>}"
}
/>
```

The following code show the use of jsp:UseBean action

```
<jsp:setPropertyname="mybean" property="*"/>
<jsp:setProperty name="mybean" property="username"/>
<jsp:setProperty name="mybean" property="username" value="Ankur"/>
```

8.9 jsp:getProperty

The jsp:getProperty element gets a bean property value using the property's getter method and inserts the value into the response.

JSP syntax The JSP syntax of jsp:get Property is given below.

```
<jsp:getProperty name="beanInstanceName" property="propertyName" />
```

The following code show the use of jsp:getProperty action

```
<jsp:useBean id="calendar" scope="page" class="emp.Calendar"/>
<h2>
Calendar of <jsp:getProperty name="calendar" property="username"/>
</h2>
```

8.10 jsp:include

The jsp:include element allows you to include either a static or dynamic resource in a JSP page. The results of including static and dynamic resources are quite different. If the resource is static, its content is included in the calling JSP page. If the resource is dynamic, it acts on a request and sends back a result that is included in the JSP page. When the include action is finished, the JSP container continues processing the remainder of the JSP page.

You cannot always determine from a path name if a resource is static or dynamic. For example, http://server:8080/index.html, might map to a servlet through a server alias. The jsp:include element handles both types of resources, so it is convenient to use when you do not know whether the resource is static or dynamic.

If the included resource is dynamic, you can use a jsp:param clause to pass the name and value of a parameter to the resource. As an example, you could pass the string username and a user's name to a login form that is coded in a JSP page.

JSP syntax The JSP syntax of jsp:include is given below.

```
<jsp:include page="{relativeURL|'${'Expression'}'|
<%= expression %>}" [flush="true|false"]
{/> |> [< jsp:param name="parameterName"
value="{parameterValue|'${'Expression'}'|
<%= expression %>}"/>]+
</jsp:include>}
```

XML syntax The XML syntax of jsp:include is given below.

```
<jsp:include page="{relativeURL|'${'Expression'}'|
%= expression %}" [flush="true|false"]
{/ >| >[ <jsp:param name="parameterName"
value="{parameterValue|'${'Expression'}'|
 %= expression %}"/>]+
</jsp:include> }
```

Attributes The attributes of jsp:include are listed below.

page="{relativeURL|<%=expression%>}" The relative URL locates the resource to be included, or an expression that evaluates to a String equivalent to the relative URL. The relative URL looks like a path name, it cannot contain a protocol name, port number, or domain name. The URL can be absolute or relative to the current JSP page. If it is absolute (beginning with a /), the path name is resolved by your web or application server.

flush="true|false" If the page output is buffered and the flush attribute is given a true value, the buffer is flushed prior to the inclusion, otherwise the buffer is not flushed. The default value for the flush attribute is false.

<jsp:paramname="parameterName'value="{parameterValue|<%= expression%>}"/>+ The jsp:param clause allows you to pass one or more name/value pairs as parameters to an included resource. The included resource should be dynamic, that is, a JSP page, servlet, or other resource that can process the parameter.

You can use more than one jsp:param clause if you want to send more than one parameter to the included resource. The name attribute specifies the parameter name and takes a case-sensitive literal string. The value attribute specifies the parameter value and takes either a case-sensitive literal string or an expression that is evaluated at request time.

Example 8.18

This example create two files date.jsp, include.jsp where include.jsp forward the request to date.jsp using forward action

Following is the content of date.jsp file

```
<p>
Today's date: <%= (newjava.util.Date()).toLocaleString()%>
</p>
Here is the content of include.jsp file
<html>
<head>
<title>The include Action Example</title>
</head>
<body>
<center>
<h2>The include action Example</h2>
<jsp:forward page="date.jsp"/>
</center>
</body>
</html>
```

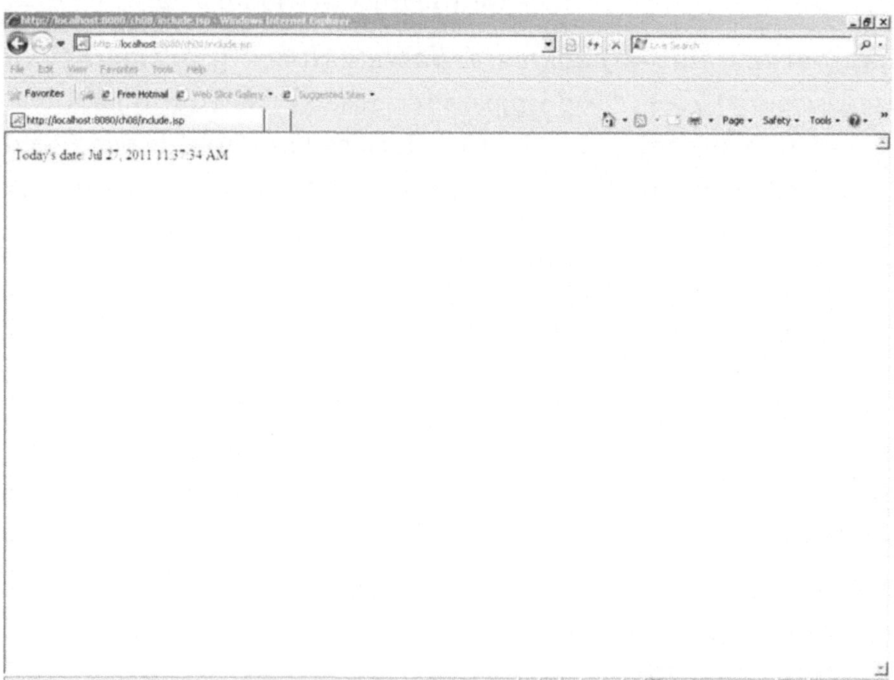

Figure 8.5 Output of include.jsp

This screen (Figure 8.5) display the result of include.jsp which forward to date.jsp and show current date and time.

Now let us keep all these files in root directory and try to access include.jsp. This would display result something like, as below. Here it discarded content from include.jsp page and displayed content from forwarded page only.

***Example* 8.9**

index.jsp

```
<%@ page errorPage="errorpage.jsp" %>
<html>
 <head>
  <title>Employee Information</title>
 </head>
 <body>
  <table width="100%" cellspacing="0">
   <tr>
    <td>
 <jsp:include page="header.jsp" flush="true">
 <jsp:param name="employee" value="Ankur"/>
  <jsp:param name="title" value="Engineer"/>
 </jsp:include>
  </td>
 </tr>
</table>
 </body>
</html>
```

header.jsp

```
<%
 out.println("< b > Employee: </b>" + request.getParameter("employee"));
 out.println("< br> <b> Title: </b>" + request.getParameter("title"));
%>
```

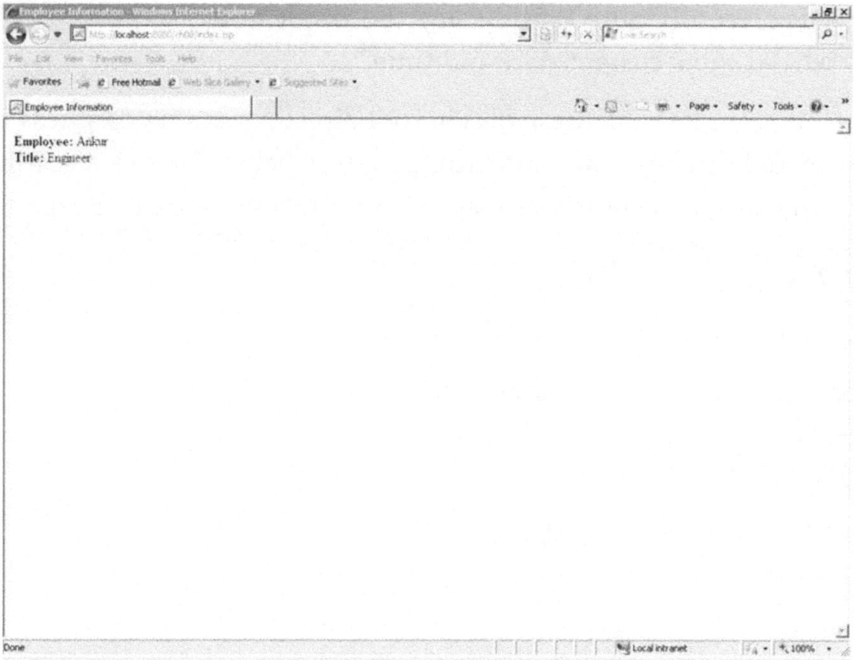

Figure 8.6 Output of index.jsp

This screen (Figure 8.6) display the value of name and title of employee.

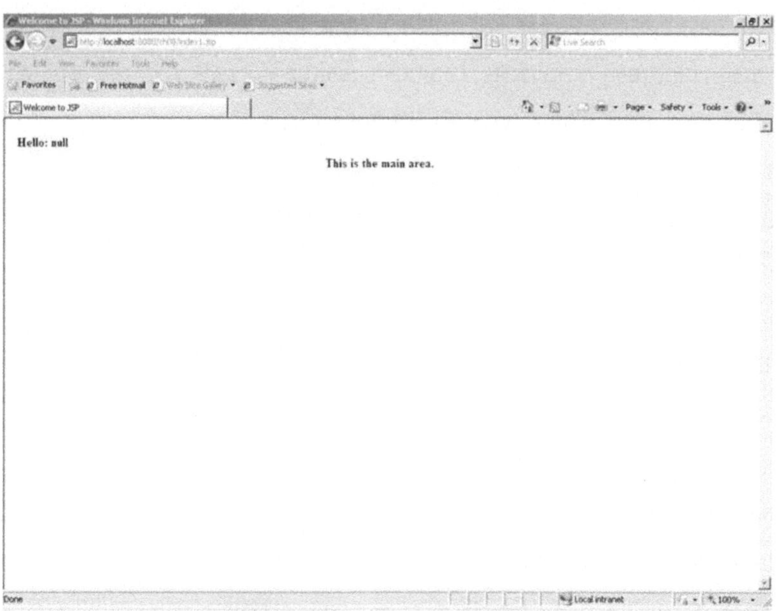

Figure 8.7 Output of index.jsp

This screen (Figure 8.7) show the result of index1.jsp under the title.jsp.

***Example* 8.10** index1.html

This example create two files index1.html ,title.jsp and display the name of user if exist using session

```
<%@ page errorPage="errorpage.jsp" %>
<html>
 <head>
  <title>Welcome to JSP</title>
 </head>
 <body>
 <table width="100%">
 <tr>
  <td>
   <%@ include file="title.jsp" %>
 </td>
 </tr>
 <tr>
  <td>
  <%
  out.println("<center> <b>This is the main area.</b> </center>");
  %>
  </td>
  </tr>
 </table>
 </body>
</html>
```

File: title.jsp

```
<table>
 <tr>
 <td>
```

```
<%
// Get the User's Name from the session
out.println("<b>Hello: " + request.getParameter("user") + "</b>");
%>
</td>
</tr>
</table>
```

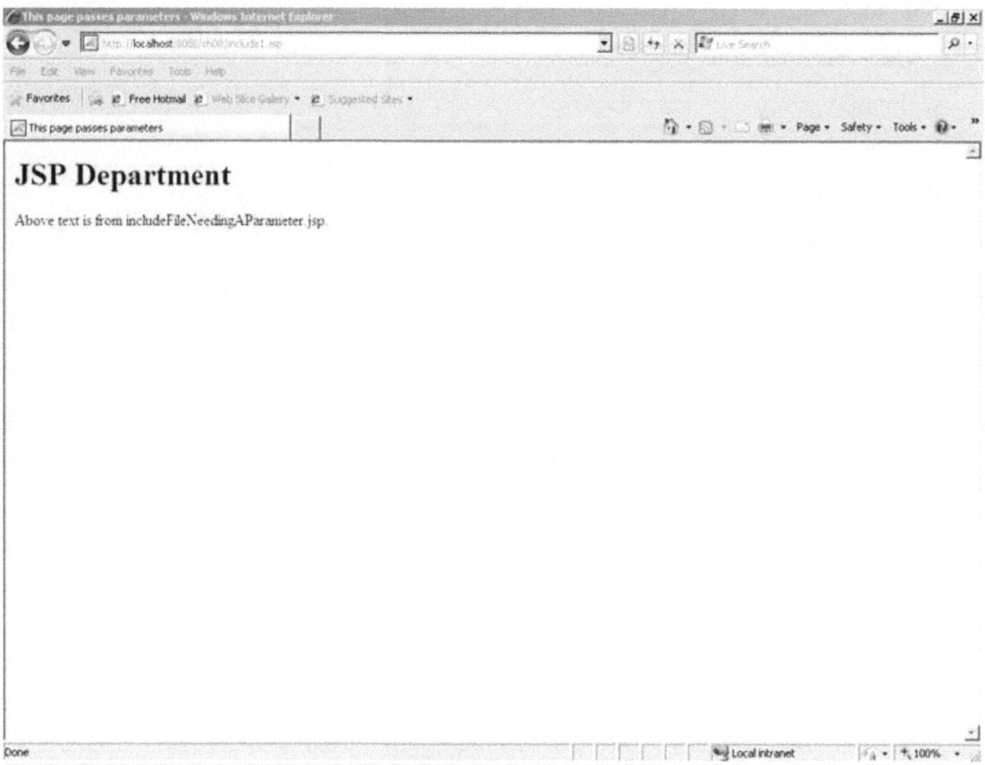

Figure 8.8 Output of include1.jsp

This screen (Figure 8.8) show the value of computer department.

***Example* 8.11**

include1.jsp

```
<html>
<head>
```

```
<title>This page passes parameters</title>
</head>
<body>
<!-- header from include-->
<jsp:include page="includeFileNeedingAParameter.jsp">
  <jsp:param name="computer" value="JSP"/>
</jsp:include>
Above text is from includeFileNeedingAParameter.jsp.
</body>
</html>
File: includeFileNeedingAParameter.jsp
<h1> <%=request.getParameter("computer") %> Department </h1>
```

8.10.1 Differences between Directive Include and **jsp: include**

`<%@ include file='filename'%>` is the JSP include directive. At JSP page translation time, the content of the file given in the include directive is 'pasted' as it is, in the place where the JSP include directive is used. Then the source JSP page is converted into a java servlet class. The included file can be a static resource or a JSP page. Generally, JSP include directive is used to include header banners and footers.

The JSP compilation procedure is that, the source JSP page gets compiled only if that page has changed. If there is a change in the included JSP file, the source JSP file will not be compiled and the modification will not reflect in the output.

`<jsp:include page="relativeURL"/>` is the JSP include action element.

The `jsp:include` action element is like a function call. At runtime, the included file will be 'executed' and the resulting content will be included with the soure JSP page. When the included JSP page is called, both the request and response objects are passed as parameters.

If there is a need to pass additional parameters, then `jsp:param` element can be used. If the resource is static, its content is inserted into the calling JSP file, since there is no processing needed.

8.11 jsp:invoke

The jsp:invoke standard action takes the name of an attribute that is a fragment, and invokes the fragment, sending the output of the result to the JspWriter, or to a scoped attribute that can be examined and manipulated. If the fragment identified by the given name is null, jsp:invoke will behave as though a fragment passed in that produces no output.

The most basic usage of this standard action will invoke a fragment with the given name with no parameters. It is also possible to invoke the fragment and send the results to a scoped attribute for further examination and manipulation. This can be accomplished by specifying the var or varReader attribute in the action. If var is specified, the container stores the result in an EL variable of type String with the name specified by var. If varReader is specified, the container stores the result in an EL variable of type java.io.Reader, with the name specified by varReader. The Reader object can then be passed to a custom tag for further processing. A translation error occurs if both var and varReader are specified.

JSP syntax The JSP syntax of jsp.invoke is given below.

```
<jsp:invoke fragment="fragmentName"
({var="scopedAttributeName"|
varReader="scopedAttributeName"}
[scope="page|request|session|application"]/>)|/>
```

Attributes The attributes of jsp:invoke are listed below.

fragment="fragmentName" The name used to identify this fragment during this tag invocation.

var="scopedAttributeName" The name of a scoped attribute to store the result of the fragment invocation in, as a java.lang.String object. A translation error must occur if both var and varReader are specified. If neither var nor varReader are specified, the result of the fragment goes directly to the JspWriter.

varReader="scopedAttributeName" The name of a scoped attribute to store the result of the fragment invocation in, as a java.io.Reader object. A translation error must occur if both var and varReader are specified. If neither var nor varReader are specified, the result of the fragment invocation goes directly to the JspWriter.

scope="page|request|session|application"

***Example* 8.12**

The following tag file represents a tag that renders the catalogue of a book database as an HTML table. The tag file declares that it sets variables, which are used by two fragment attributes. Before the tag invokes the fragment attributes using the jsp:invoke element, the web container passes values for the variables back to the calling page.

```
<%@ attribute name="jsp" required="true"
 type="java.jsp" %>
<%@ attribute name="color" required="true"%>
<%@ attribute name="normalPrice" fragment="true"%>
<%@ attribute name="onSale" fragment="true"%>
<%@ variable name-given="price"%>
<%@ variable name-given="salePrice" %>
<center>
<table>
<c:forEach var="book" begin="0" items="${bookjava.books}">
 ...
<c:set var="salePrice" value="${book.price *.85}"/>
 <c:set var="price" value="${book.price}"/>
 <c:choose>
  <c:when test="${book.onSale}">
   <jsp:invoke fragment="onSale"/>
  </c:when>
  <c:otherwise>
   <jsp:invoke fragment="normalPrice"/>
  </c:otherwise>
 </c:choose>
...
</table>
</center>
```

8.12 jsp:output

Specifies the XML declaration or the document type declaration in the request output of a JSP document or a tag file that is in XML syntax.

The jsp:output element specifies the XML Declaration or the Document Type Declaration in the request output of the JSP document.

The XML Declaration and Document Type Declaration that are declared by the jsp:output element are not interpreted by the JSP container. Instead, the container simply directs them to the request output.

8.12.1 Generating a Document Type Declaration

A Document Type Declaration (DTD) defines the structural rules for the XML document in which the Document Type Declaration occurs. XML documents are not required to have a DTD associated with them.

Specifying the Document Type Declaration in the jsp:output element will not cause the JSP container to validate the JSP document against the DTD.

If you want the JSP document to be validated against the DTD, you must manually include the Document Type Declaration within the JSP document, just as you would with any XML document.

8.12.2 Generating XML Declarations

Here is an example of an XML declaration

```
<?xmlversion="1.0"encoding="UTF-8"?>
```

This declaration is the default XML declaration. It means that if the JSP container is generating an XML declaration, this is what the JSP container will include in the output of your JSP document.

Neither a JSP document nor its request output is required to have an XML declaration. In fact, if the JSP document is not producing XML output then it should not have an XML declaration.

The JSP container will not include the XML declaration in the output when either of the following is true.

- You set the omit-xml-declaration attribute of the jsp:output element to either true or yes.
- You have a jsp:root element in your JSP document, and you do not specify omit-xml-declaration="false" in jsp:output.

The JSP container will include the XML declaration in the output when either of the following is true.

- You set the omit-xml-declaration attribute of the jsp:output element to either false or no.
- You do not have a jsp:root action in your JSP document, and you do not specify the omit-xml-declaration attribute in jsp:output.

The jsp:output element has three attributes that you use to generate the Document Type Declaration.

- doctype-root-element Indicates the root element of the XML document
- doctype-system Indicates the URI reference to the DTD
- doctype-public A more flexible way to reference the DTD. This identifier gives more information about the DTD without giving a specific location. A public identifier resolves to the same actual document on any system even though the location of that document on each system may vary.

The rules for using the attributes are as follows.

- The doctype attributes can appear in any order
- The doctype-root attribute must be specified if the doctype-system attribute is specified
- The doctype-public attribute must not be specified unless doctype-system is specified

XML syntax The XML syntax for jsp:output given below.

```
<jsp:output(omit-xml-declaration="yes|no|true|false")
{ doctypeDecl}/>
  doctypeDecl ::= (doctype-root-element="rootElement"
  doctype-public="PubidLiteral"
```

```
doctype-system="SystemLiteral")
|(doctype-root-element="rootElement"
  doctype-system="SystemLiteral")
```

Attributes The attributes of jsp:output are listed below.

omit-xml-declaration="yes|no|true|false" Indicates whether to omit the generation of an XML declaration.

doctype-root-element="rootElement" Must be specified if and only if doctype-system is specified or a translation error must occur. Indicates the name that is to be output in the generated DOCTYPE declaration.

doctype-system="SystemLiteral" Specifies that a DOCTYPE declaration is to be generated and gives the value for the System Literal.

doctype-public="PubidLiteral" Must not be specified unless doctype-system is specified. Gives the value for the Public ID for the generated DOCTYPE.

***Example* 8.13**

Here is an example of specifying a Document Type Declaration with jsp:output.

```
<jsp:output doctype-root-element="books"
  doctype-system="books.dtd" />
```

The resulting output is

```
<!DOCTYPE books SYSTEM "books.dtd" >
```

8.13 jsp:plugin

Causes the execution of an applet or bean. The applet or bean executes in the specified plugin. If the plugin is not available, displays a dialog to initiate the download of the plugin software.

The jsp:plugin element plays or displays an object (typically an applet or bean) in the client web browser, using a Java plug-in that is built in to the browser or downloaded from a specified URL.

When the JSP page is translated and compiled and sends back an HTML response to the client, the jsp:plugin element is replaced by either an

<object>or<embed> element, according to the browser version. The <object> element is defined in HTML.

In general, the attributes to the jsp:plugin element specify whether the object is a bean or an applet, locate the code that will be run, and position the object in the browser window, specify an URL from which to download the plug-in software, and pass parameter names and values to the object. The attributes are described in detail in the next section.

JSP syntax The JSP syntax of jsp:plugin is given below.

```
<jsp:plugin
  type="bean|applet"
  code="classFileName"
  codebase="classFileDirectoryName"
  [name="instanceName"]
  [archive="URIToArchive, ..."]
  [align="bottom|top|middle|left|right"]
  [height="{displayPixels|<%= expression %>}"]
  [width="{displayPixels|<%= expression %>}"]
  [hspace="leftRightPixels"]
  [vspace="topBottomPixels"]
  [jreversion="JREVersionNumber|1.2"]
  [nspluginurl="URLToPlugin"]
  [iepluginurl="URLToPlugin"]>
  [<jsp:params>
    [<jsp:param name="parameterName"
      value="{parameterValue|'${'Expression'}'
 |<%= expression %>}"/>]+ </jsp:params>]
      [<jsp:fallback> text message if plugin download fails
    </jsp:fallback>]
    </jsp:plugin>
```

XML syntax The XML syntax of jsp:plugin is given below.

```
<jsp:plugin
type="bean|applet" code="classFileName"
codebase="classFileDirectoryName"
[name="instanceName"][archive="URIToArchive, ..."]
[align="bottom|top|middle|left|right"]
[height="{displayPixels|'${'Expression'}'|%= expression %}"]
[width="{displayPixels|'${'Expression'}'|%= expression %}"]
[hspace="leftRightPixels"][vspace="topBottomPixels"]
   [jreversion="JREVersionNumber|1.2"]
[nspluginurl="URLToPlugin"]
[iepluginurl="URLToPlugin"]>
[<jsp:params>
[<jsp:param name="parameterName"
value="{parameterValue|'${'Expression'}'|
%= expression %}"/>]+
</jsp:params>]
[<jsp:fallback>textmessage if plugin download fails
   </jsp:fallback>]
</jsp:plugin>
```

Attributes The attributes of jsp:plugin are listed below.

type="bean|applet" The type of object the plug-in will execute. You must specify either bean or applet, as this attribute has no default value.

code="classFileName" The name of the Java class file the plug-in will execute. You must include the .class extension in the name. The class file you specify should be in the directory named in the codebase attribute.

codebase="classFileDirectoryName" The directory (or path to the directory) that contains the Java class file the plug-in will execute. If you do not supply a value, the path of the JSP page that calls jsp:plugin is used.

name="instanceName" A name for the instance of the bean or applet, which makes it possible for applets or Beans called by the same JSP page to communicate with each other.

archive="URIToArchive, ..." A comma-separated list of pathnames that locate archive files that will be preloaded with a class loader located in the directory named in codebase. The archive files are loaded securely, often over a network, and typically improve the applet's performance.

align="bottom|top|middle|left|right" Specifies the position of the image, object, or applet. The position descriptions listed below use the term text line to mean the line in the viewable JSP page that corresponds to the line in the JSP page where the jsp:plugin element appears. The allowed values for align are listed below.

bottom Aligns the bottom of the image with the baseline of the text line.

top Aligns the top of the image with the top of the text line.

middle Aligns the vertical centre of the image with the baseline of the text line.

left Floats the image to the left margin and flows text along the image's right side.

right Floats the image to the right margin and flows text along the image's left side.

height="{displayPixels|<%= expression %>}"
width="{displayPixels|<%= expression %>}"

The initial height and width, in pixels of the image, the applet or bean displays not counting any windows or dialog boxes the applet or bean brings up.

hspace="leftRightPixels" vspace="topBottomPixels" The amount of space, in pixels, to the left and right (or top and bottom) of the image the applet or bean displays. The value must be a non-zero number. Note that hspace creates space to both the left and right and vspace creates space to both the top and bottom.

jreversion="JREVersionNumber|1.2" The version of the Java Runtime Environment (JRE), the applet or bean requires. The default value is 1.2.

nspluginurl="URLToPlugin" The URL is where the user can download the JRE plug-in for Netscape Navigator. The value is a full URL, with a protocol name, optional port number, and domain name.

iepluginurl="URLToPlugin" The URL where the user can download the JRE plug-in for Internet Explorer. The value is a full URL, with a protocol name, optional port number, and domain name.

<jsp:params>[<jsp:paramname="parameterName"value="{parameterValue| <%=expression%>}"/>]+</jsp:params> The parameters and values that you want to pass to the applet or bean. To specify more than one parameter value, you can use more than one jsp:param element within the jsp:params element. The name attribute specifies the parameter name and takes a case-sensitive literal string. The value attribute specifies the parameter value and takes either a case-sensitive literal string or an expression that is evaluated at runtime. If the dynamic resource you are passing the parameter to isan applet, it reads the parameter with the java.applet.Applet.getParameter method.

<jsp:fallback> text message for user </jsp:fallback> A text message to display for the user if the plug-in cannot be started. If the plug-in starts but the applet or bean does not, the plug-in usually displays a popup window explaining the error to the user.

The following code display the usage of plugin action

```
<jsp:plugintype=applet code="Molecule.class" codebase=" html">
<jsp:params>
<jsp:param name="molecule" value="molecules/jsp.mol"/>   </jsp:params>
      <jsp:fallback>
    <p>Unable to load applet</p>
  </jsp:fallback>
</jsp:plugin>
```

8.14 jsp:root

Defines standard JSP elements and namespace attributes of tag libraries. A JSP page in XML syntax can have jsp:root as its root element, but it is not required. You can instead specify your own tag as a root element. Tag libraries used within the JSP page can be represented in the root element through xmlns attributes. Again, you can use the xmlns attributes in other tags besides the jsp:root element. The xmlns:jsp and version attributes are mandatory.

XML syntax The XML syntax of jsp:root is given below.

```
<jsp:root
```

```
  xmlns:jsp="http://java.sun.com/JSP/Page"
  [xmlns:taglibPrefix="URI" ]+ ...
  version="1.2|2.0">
    JSP Page
</jsp:root>
```

Attributes The Attributes of jsp:root are listed below.

xmlns:jsp="http://java.sun.com/JSP/Page" Enables the use of the standard elements defined in the JSP specification.

version="1.2|2.0" Specifies the version of the JSP specification the page is using.

xmlns:taglibPrefix="URI"

8.15 jsp text

A jsp:text element is used to enclose template data in the XML representation. A jsp:text element has no attributes and can appear anywhere that template data can. The interpretation of a jsp:text element is to pass its content through to the current value of out.

XML syntax allows an XML element that does not represent a standard or custom action to appear anywhere a jsp:text can appear. Such an element is passed to the current out.

JSP syntax The JSP syntax of jsp:text is given below.

```
<jsp:text>
  templatedata
</jsp:text>
```

XML syntax The XML syntax of jsp:text is given below.

```
<jsp:text>
  template data
</jsp:text>
```

***Example* 8.14**

```
<hello> <jsp:scriptlet>i=3;</jsp:scriptlet>
```

```
<hi>
<jsp:text> hi Ankur Saxena
</jsp:text>  <jsp:expression> i< /jsp:expression>
</hi>
</hello>
```

8.16 jsp:param

JSP Include Param is used to pass parameter in the include directive. The example uses <jsp:param> clause to pass the parameter name and value to the dynamic jsp page. This clause is only used with the <jsp:include> directive. It allows you to include either a static or dynamic file in a JSP page. The <jsp:include> acts on a request and sends back the result included in the JSP page.

Example 8.15

include2.jsp

```
<html>
<body>
<font color="green"> <h2>Login Information</h2> </font>
<jsp:include page="includeparam.jsp" flush="true">
<jsp:param name="username" value="Ankur"/>
<jsp:param name="password" value="Saxena"/>
</jsp:include>
</body>
</html>
```

includedparam.jsp

```
<html>
<%
out.println("<b>Username:</b>"+request.getParameter("username")+"<br>");
out.println(" <b> Password: </b>" + request.getParameter("password"));
%>
</html>
```

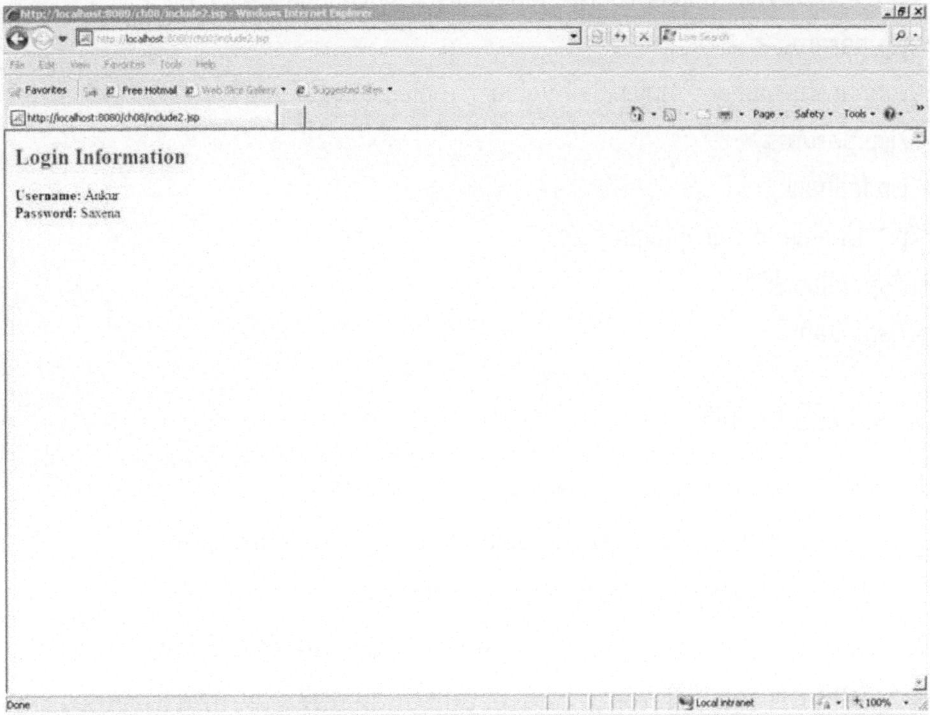

Figure 8.9 Output of include 2.jsp

This screen (Figure 8.9) display the name and password using param action

8.17 jsp:fallback

The <jsp: fallback> tag indicates the content to be used by the client browser if the plugin cannot be started, either because OBJECT or EMBED is not supported by the client browser, or due to some other problem. If the plugin can start but the Applet or JavaBeans component cannot be found or started, a plugin specific message will be presented. Usually the message is a pop-up window reporting a ClassNotFoundException.

Syntax The syntax of jsp:fallback is given below.

```
<jsp:fallback>
 Text message for user
</jsp:fallback>
```

The following code display the usage of fallback action

```
<jsp:plugin type=applet code="Molecule.class codebase="/html">
<jsp:params>
<jsp:param name="molecule" value="molecules/jsp.mol"/>
</jsp:params>
<jsp:fallback>
<p> Unable to start plugin </p>
</jsp:fallback>
</jsp:plugin>
```

REVIEW QUESTIONS

1. What are the main features of jsp actions?
2. What is the role of jsp:attribute action in EL expression.
3. How to merge jsp: body with jsp: element.
4. What is the difference between jsp:body and jsp:doBody actions
5. What is the difference between include directive and jsp:include action.
6. What is the difference between jsp:forward and jsp:include action.
7. What is jsp: invoke and jsp: output action
8. What are jsp and xml syntax of jsp: text action
9. What is the concept of jsp: Param action? Explain with suitable example.

9

JSP WITH IMPLICIT OBJECTS

9.1 IMPLICIT OBJECTS

JSP Scripting elements provide a great deal of power for creating, modifying and interacting with Java objects in order to generate dynamic content. Application-specific classes can be instantiated and values from method calls can be accessed to store and retrieve data for use by JSP pages.

In addition to objects, which are completely under the control of developer, the JSP container also exposes a number of its internal objects to the page author. These are referred to as implicit objects, because their availability in a JSP page is automatic. The developer can assume that these objects are present and accessible via JSP scripting elements. The objects will be automatically assigned to specific variable names in the page's scripting language. As described in the Table 9.1, each implicit object must adhere to a corresponding API, in the form of a specific java class of interface definition.

The nine implicit objects provided by JSP fall naturally into four main categories: objects related to a JSP page's servlet, objects concerned with page input and output, objects providing information about the context within which a JSP page is being processed, and objects resulting from errors. Beyond this functional categorization, four of the JSP implicit objects-request session, application, and page context have something else in common the ability to store and retrieve arbitrary attribute values. By setting and getting attribute values, these objects are able to transfer information between and among JSP pages and servlets as a simple data-sharing mechanism.

Table 9.1 Implict Objects

Variable	Class	Description
application	javax.servlet.ServletContext	The context for the JSP page's servlet and any web components contained in the same application.
Config	javax.servlet.ServletConfig	Initialization information for the JSP page's servlet.
Exception	java.lang.Throwable	Accessible only from an error page.
Out	javax.servlet.jsp.JspWriter	The output stream.
Page	java.lang.Object	The instance of the JSP page's servlet processing the current request.
pageContext	javax.servlet.jsp.PageContext	The context for the JSP page. Provides a single API to manage the various scoped attributes . This API is used extensively when implementing tag handlers.
Request	Subtype of javax.servlet.ServletRequest	The request triggering the execution of the JSP page.
Response	Subtype of javax.servlet.ServletResponse	The response to be returned to the client.
Session	javax.servlet.http.HttpSession	The session object for the client.

The standard method for attribute management provided by the classes and interfaces of these four objects are summarized in the table below. Note that attribute keys take the form of Java String Objects, while their values are referred as instances of java.lang.Object.

Common method for storing and retrieving attribute values. are given in Table 9.2

Table 9.2 Methods for storing and Retrievins Attribute Values

Method	**Description**
setAttribute (key,value)	Associate an attribute value with a key.
getAttributeNames ( )	Retrieves the names of all attributes associated with session.
getAttribute (key)	Retrieves the attribute values associated with a key.
removeAttribute (key)	Removes the attribute values associated with a key

9.2 SERVLET-RELATED OBJECTS

The objects in this category are based on the JSP page's implementations as a servelet. The page implicit object represents the servlet itself, while the config object stores the servlet' s initialization parameters, if any.

9.2.1 Page Object

The page object represents the jsp page itself or more specifically an instance of the servlet class into which the page has been translated. It may be used to call any of the methods defined by the servlet class. The extends attribute of the page directive may be used to specify the servlet class explicity, otherwise an implement-specific class will be used by the JSP container when constructing the servlet. In either case ,the servlet class is always required to implement the javax.servlet.jsp.Jsp Page interface. In the specific case of web-based JSP application built on HTTP, the servlet class must implement the javax.servlet.jsp.HttpJspPage interface. Here is an example page fragment that utilizes this implicit object.

```
<%@ page info="page implicit oblect demo" %>
Page info:
<%=(( javax.servlet.jsp.HttpJspPage)page).getServletInfo() %>
```

This expression will insert the value of the page's documentation string into the output from the page. In this example, note that because the servlet class varies from one page to another, the standard type for the page implicit object is the default java type of non-primitive values, java.lang.Object.

This object has a page scope and is an instance of the JSP page's servlet class that processes the current request. Page object represents the current page that is used to call the methods defined by the translated servlet class. First, type -cast the servlet before accessing any method of the servlet through the page.

The page implicit object is of type Object and it is assigned a reference to the servlet executing the jspService() method. Thus in the Servlet generated by Tomcat the page object is created as

```
Object page = this;
```

Because page is of type Object it is of less direct use than simply accessing the implicit 'this' object. To access any of the methods of the servlet through page it must be first cast to type Servlet. Thus the two following lines of code are equivalent.

```
<% this.log("log message"); %>
<% ((HttpServlet)page).log("anothermessage"); %>
```

9.2.2 Config Object

Config object is related to get servlet's configuration information. It is in javax.servlet.ServletConfig interface. Servlet configuration object is used to pass information of servlet to servlet container during initialization of servlet. This is defined in web.xml file with <init-param> and method is used to get getServletConfig()

The following code show the usage of congig object

```
<servlet>
  <servlet-name>ServletName</servlet-name>
  <servlet-class>com.myapp.servlet.ServletName</servlet-class>
    <init-param>
    <param-name>dbUserName</param-name>
    <param-value>root</param-value>
  </init-param>
</servlet>
```

Method used to get init param is getInitParameter(), to get value inside <param-value> and getInitParameterNames() to get name of <param-name>

The config object can be used to retrieve parameters using its getInitParameter(String param) method.

Apache Tomcat

```
<%
ServletContext ct = config.getServletContext();
out.print(ct.getServerInfo());
%>
```

Get the value of a configuration parameter.

```
<%
String param1 = config.getInitParameter("param1");
%>
```

Get the username and password

```
<%! static private DBConnectionPool pool=null;
public void jspInit()
{
if(pool==null)
{
String username=config.getInitParameter("username");
String password= config.getInitParameter("password");
pool= DBConnectionPool.getPool(this,username,password);
}
}
%>
```

9.3 INPUT/OUTPUT OBJECTS

These objects are focused on the input and output of a JSP page. The request object represents the data coming into the page, while the response object represents its result. The out implicit object represents the actual output stream associated with the response, to which page the page's content is written.

9.3.1 Request object

The request object retrieves the values that the client browser passed to the server during an HTTP request such as headers, cookies or parameters associated with the request. Among the most common use of the request object is to obtain parameter or query string values. The following demonstration illustrates how to use the request object for parsing form data.

Request object has a request scope that is used to access the HTTP request data, and also provides a context to associate the request-specific data. Request object implements javax.servlet.ServletRequest interface. It uses the getParameter() method to access the request parameter. The container passes this object to the jspService() method.

The following table summaries the most useful methods available to the request object.

Method of the javax.servlet.http.HttpServletRequest interface for accessing request parameters are given in Table 9.3

Table 9.3 Methods for accessing request Parameter

Method	Description
getParameter(String name)	Returns the value of a request parameter as a String, or null, if the parameter does not exist.
getParameterNames()	Returns an Enumeration of String object containing the names of the parameters contained in this request.
getParameterValues(String name)	Returns an array of String objects containing all of the values the given request parameter has, or null if the parameter does not exist.

Method of the javax.servlet.http.HttpServletRequest interface for retrieving request headers are given in Table 9.4

Table 9.4 Methods for retrievies request headers

Method	Description
getHeaderNames()	Returns an enumerator of all HTTP header names.
getHeader(String hdr)	Returns the value of an HTTP header such as "QUERY_STRING" or "URL". The method getHeaderNames() can be used to determine what headers are available. See example below.
getHeaders(name)	Returns all of the values for a single request header.
getIntHeader(name)	Returns the values of a single request header as in integer
getDateHeader(name)	Returns the values of a single request header as a date.
getCookies()	Gets the array of cookies found in this request. See below for more details on working with cookies.

Miscellaneous Method of the javax.servlet.http.HttpServletRequest interface are given in Table 9.5

Table 9.5 Miscellaneous Methods

Method	Description
getMethod()	Returns the HTTP(e.g. GET,POST)method for the request.
getRequestURL()	Gets the URL to the current JSP page.
getSession(flag)	Retrieves the session data for the request.
getQueryString()	Gets any query string that is part of the HTTP request URL.
getRequestDispatcher(path)	Creates the request dispatcher for the indicated local URL.
getRemoteHost()	Returns the fully qualified name of the host that sent the request.
getRemoteAddr()	Returns the network address of the host that sent request.
getAttribute(String)	Retrieves the object associated with the named attribute.

Table 9.5 (Continues)

Method	Description
getAttributeNames()	Retrieves the names of all attributes currently associated with the session.
setAttribute(String, object)	Sets the object to the named attribute. attribute created if it doesnot exist.
removeAttribute(String)	Removes the object bound with the specified name from this session.
getRemoteUser()	Returns the name of user that sent request, if known

Print value of a named request parameter using expression.

```
<%= request.getParameter("name") %>
```

Access the value of a request attribute of type String inside scriptlet.

```
<%
 String name = request.getAttribute("name");
%>
```

Example 9.1

We will demonstrate a request object in a simple form. The form is used to send user data to the server for processing.

request.jsp

```
<html>
<head>
<title>Form</title>
<style>
* { font-size: 14px; font-family: Verdana }
input { border: 1px solid #ccc }
```

```
</style>
</head>
<body>
<form method="get" action="book.jsp">
<p>
Enter a book name
</p>
<input type="text" name="bookname">
<input type="submit" value="submit">
</form>
</body>
</html>
```

book.jsp

```
<html>
<head>
<title>Book</title>
<style>
* { font-size: 12px; font-family: verdana }
</style>
</head>
<body>
<p>
You have entered: <%= request.getParameter("bookname") %>
</p>
</body>
</html>
```

Figure 9.1 Output of request.jsp

This screen (Figure 9.1) display the book name and this page action to book.jsp

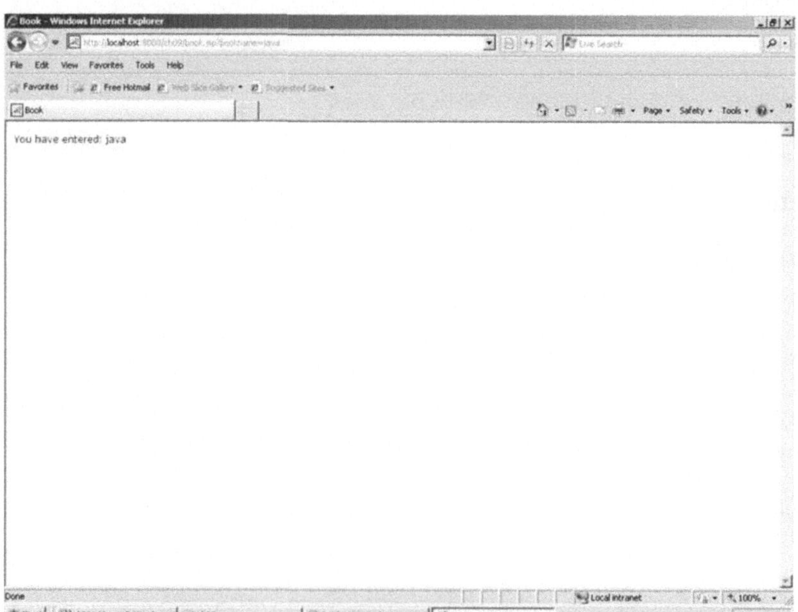

Figure 9.2 Output of book.jsp

This screen (Figure 9.2) display the name of the book entered by request.jsp page.

In this example, we have two jsp files. One file is used to enter data. We use a form. The data is sent to the second jsp file, which will simply print the data we have entered.

```
<style>
* {font-size:12px;font-family:Verdana}
input {border:1px solid #ccc}
</style>
```

We use some css to make our page prettier. We can change the font size and family, and set a new border for an input html tag.

```
<form method="get" action="book.jsp">
```

The action parameter specifies the name of the jsp file, which will receive the data we send.

```
You have entered: <%= request.getParameter("bookname") %>
```

Example 9.2

This example display the header information using request object

header1.jsp

```
<%@ page import="java.io.*,java.util.*" %>
<html>
<head>
<title>HTTP Header Request Example</title>
</head>
<body>
<center>
<h2>HTTP Header Request Example</h2>
<table width="100%" border="1" align="center">
<tr bgcolor="#99494">
<th>Header Name</th><th>Header Value(s)</th>
```

```
</tr>
<%
  Enumeration headerNames = request.getHeaderNames();
  while(headerNames.hasMoreElements()) {
    String paramName = (String)headerNames.nextElement();
    out.print("<tr><td>" + paramName + "</td>\n");
    String paramValue = request.getHeader(paramName);
    out.println("<td> " + paramValue + "</td></tr>\n");
  }
%>
</table>
</center>
</body>
</html>
```

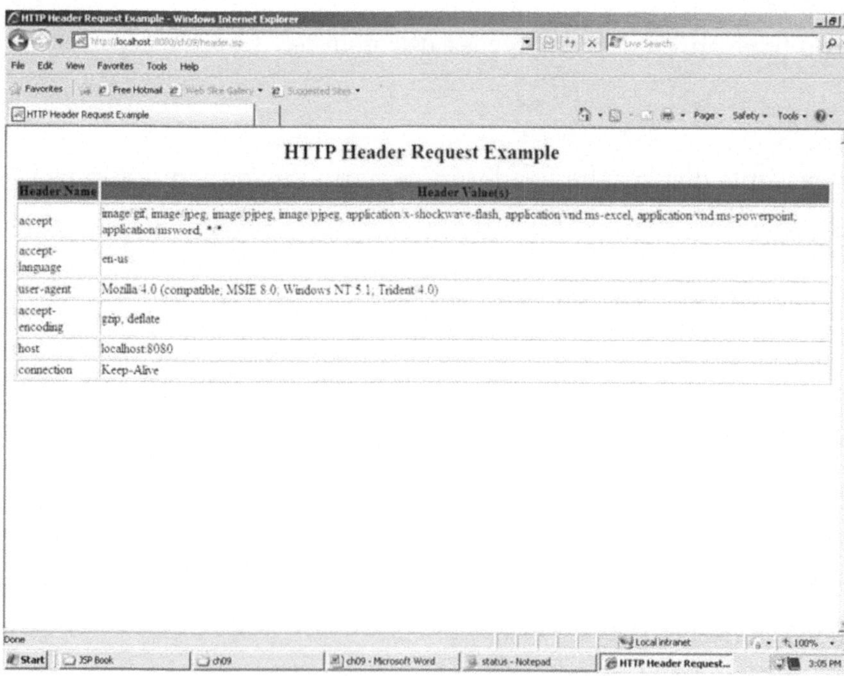

Figure 9.3 Output of header.jsp

This screen (Figure 9.3) display the header information.

9.3.2 Response Object

This object has a page scope that allows direct access to the HTTPServletResponse class object. Response object is an instance of the classes that implements the javax.servlet.ServletResponse class. Container generates to this object and passes to the jspService() method as a parameter.

The response object has page scope. There is one situation, where the response object might be useful. Say we want to rename the url of one page. But we realize that many people might have bookmarked the page. To overcome this issue, we might use the response object to redirect the request to a new page.

The response object handles the output of the client. This object can be used for creating HTTP Headers, creating cookies, setting content type and redirecting workflow.

9.3.3 Servlet Class

Method of the javax.servlet.http.HttpServletResponse interface for specifying content are shown in Table 9.6

Table 9.6 Method for Specifying Content

Method	Description
setContentType()	Sets the MIME type and character encoding for the page.
getCharacterEncoding()	Returns the character encoding style set for the response's content

Method of the javax.servlet.http.HttpServletResponse interface for response headers is shown in Table 9.7

Table 9.7 Methods for Response headers.

Method	Description
addCookie(cookie)	Adds the specified cookie to the response. It can be called multiple times to set more than one cookie.
containsHeader(name)	Checks whether the response already includes this header.
setHeader(name,value)	Creates an HTTP Header.
setIntHeader(name,value)	Assigns a specified integer value to the named header
setDateHeader(name,value)	Assigns a specified date value to the named header
addHeader(name,value)	Adds the specified string as a value to the named header
addIntHeader(name,value)	Adds the specified integer as a value to the named header
addDateHeader(name,value)	Adds the specified date as a value to the named header

Response code method of the javax.servlet.http.HttpServletResponse interface is given in Table 9.8

Table 9.8 Response Code Methods

Method	Description
setStatus(code)	Sets the status code for the response
sendError(status,msg)	Sets the status code and error message for the response
sendRedirect(String)	Sends a temporary redirect response to the client using the specified redirect location URL.
encodeRedirectURL(url)	Encodes the URL for use with the sendRedirect() method to include session information.
encodeURL(name)	Encodes an used link to include session information.

The method sendRedirect() is useful for controlling the workflow of a web application. This is a very important and often used method and so let us look a bit more closely at how this works.

The sendRedirect() method works not by redirecting the workflow on the server but by instructing the client browser to make a request for the page defined in the method. Let us walk through an example to see more closely how this works.

Consider the example of a user logging in to a web application. First, the user is displayed page containing an HTML form in which the user can enter a password. When the form is submmited the browser makes a request to the JSP defined in the action attribute of the form and attaches the password to the request.This JSP can pick up the password using request.getParameter() and then checks to see if the password is valid. If the password is not valid then the JSP makes a call to the sendRedirect() method providing the URL of the login page. As this page executes, the JVM executes the sendRedirect() method. When it does this it stops the execution of the page, attaches an HTTP header which instructs the client to redirect to the login page and serves the result to the client. The client process the HTTP headers and without attempting to display any HTML it automatically makes a request to the server for the login page. The following sequence diagram summarises this workflow.

When a browser encounters an HTTP header instructing it to immediately request another page. It leaves the whatever page is currently being displayed on screen and executes this second request in the background, so to the user the whole process looks quite seemless.

There are two important points to be aware of when using the sendRedirect() method. First of all, the method will fail if the executing JSP has already sent some page content to the browser already. When a web server sends a page to the client it does so in chunks rather than one large file. So, if one or more of these chunks have been sent to the browser then a call to sendRedirect() will fail. To avoid this scenario, you can either increase the size of the chunks by using page buffer directive or preferrably structure your JSP such that the call to sendRedirect() is placed before the opening <HTML> tag.

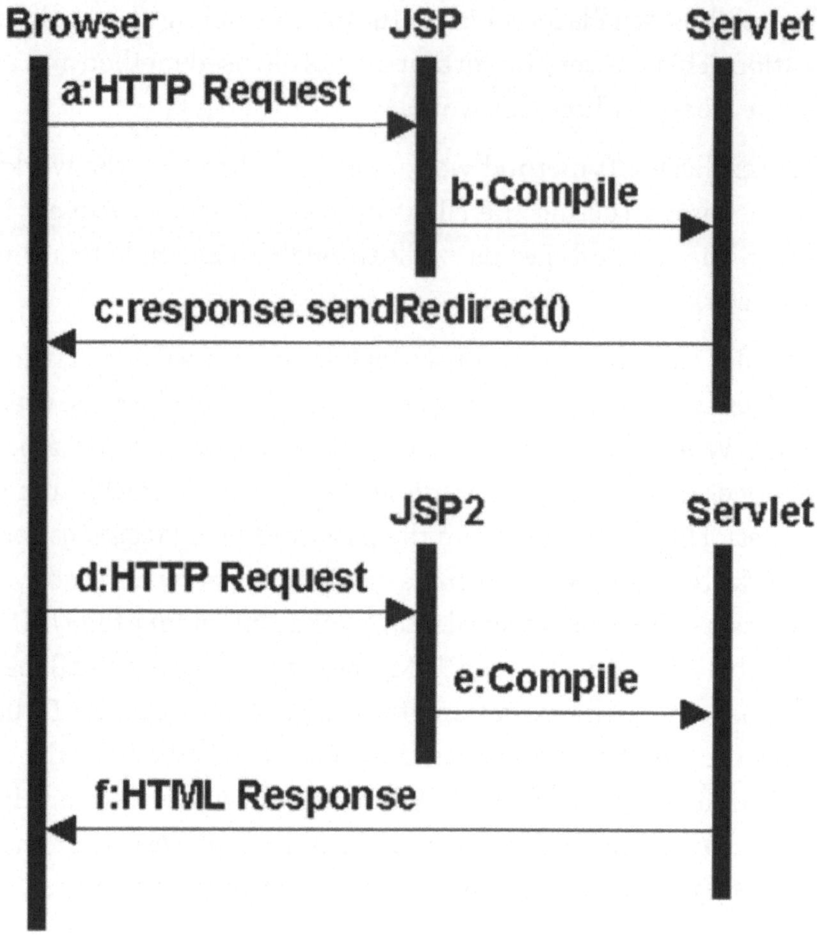

Figure 9.4 Sequence diagram for the response.sendRedirect() method

Another point is about HTTP headers. Although no page content is displayed, when the sendRedirect() method is executed, HTTP headers are processed. This means that it is possible to write code that will create a cookie and then perform a redirect. This makes sense. The code that creates a cookie does not actually create the cookie it just adds an instruction to the HTTP header that instructs the client to create cookies. Likewise, the sendRedirect() method does not actually redirect but instead it adds an instruction to the HTTP header that this time instructs the client to request another JSP.

```
<%
response.setHeader("Cache-Control", "no-cache");
%>
```

This example show redirect response for another page

index.jsp

```
<%
  response.sendRedirect("response.jsp");
%>
```

***Example* 9.3**

response.jsp

```
<%@page contentType="text/html" pageEncoding="UTF-8"%>
<html>
<head>
<title>Response JSP Page</title>
</head>
 <body>
    <h2>Welcome</h2>
  </body>
</html>
```

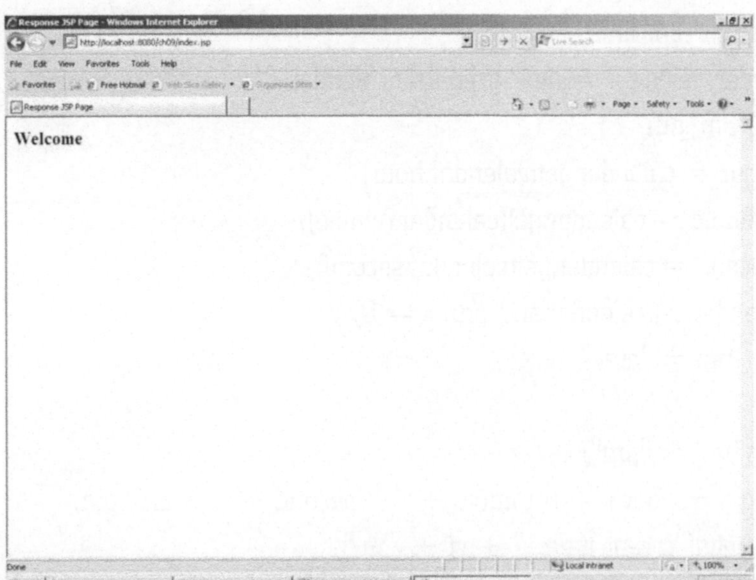

Figure 9.5 Output of index.jsp

This screen (Figure 9.5) display the result of response.jsp redirect by index.jsp

Example 9.4

This example display some header information using response object

header1.jsp

```
<%@ page import="java.io.*,java.util.*" %>
<html>
<head>
<title>refresh header example</title>
</head>
<body>
<center>
<h2>refresh header example</h2>
<%
  // set refresh, autoload time as 5 seconds
   response.setintheader("refresh", 5);
  // get current time
  calendar calendar = new gregoriancalendar();
  string am_pm;
  int hour = calendar.get(calendar.hour);
  int minute = calendar.get(calendar.minute);
  int second = calendar.get(calendar.second);
  if(calendar.get(calendar.am_pm) == 0)
     am_pm = "am";
  else
     am_pm = "pm";
  string ct = hour+":"+ minute +":"+ second +" "+ am_pm;
  out.println("crrent time: " + ct + "\n");
%>
</center>
</body>
</html>
```

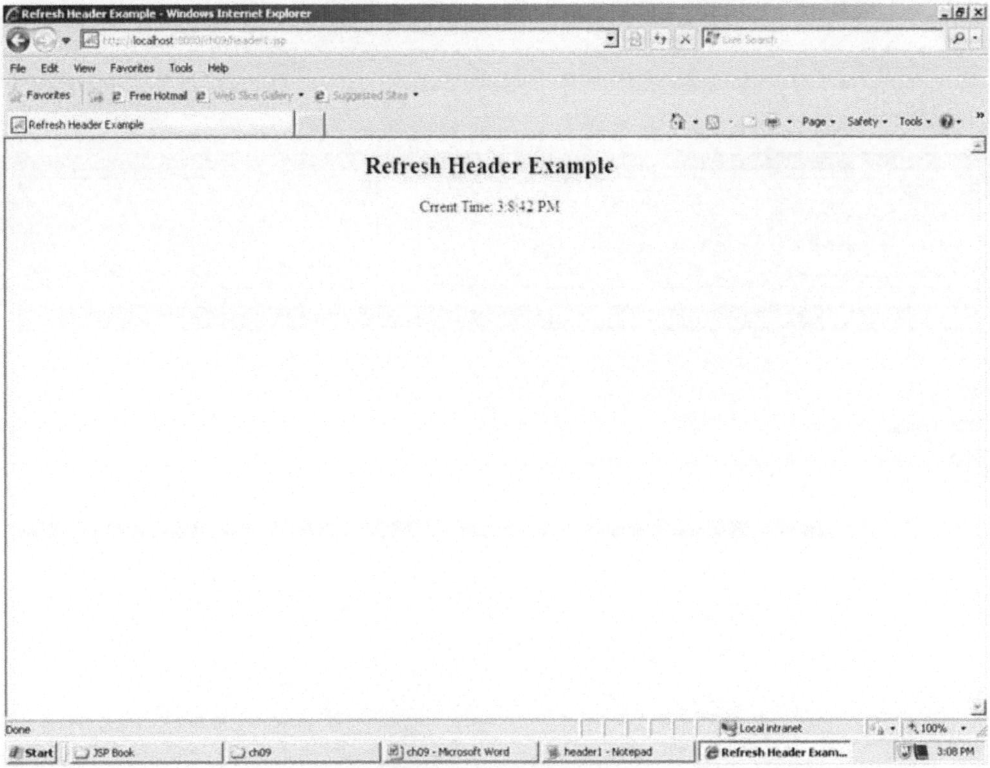

Figure 9.6 Output of header1.jsp

This screen (Figure 9.6) display refresh time

***Example* 9.5**

This example show how to set http status code

status.jsp

```
<html>
<head>
<title>setting http status code</title>
</head>
<body>
<%
   // set error code and reason.
   response.senderror(407, "need authentication!!!" );
```

```
%>
</body>
</html>
```

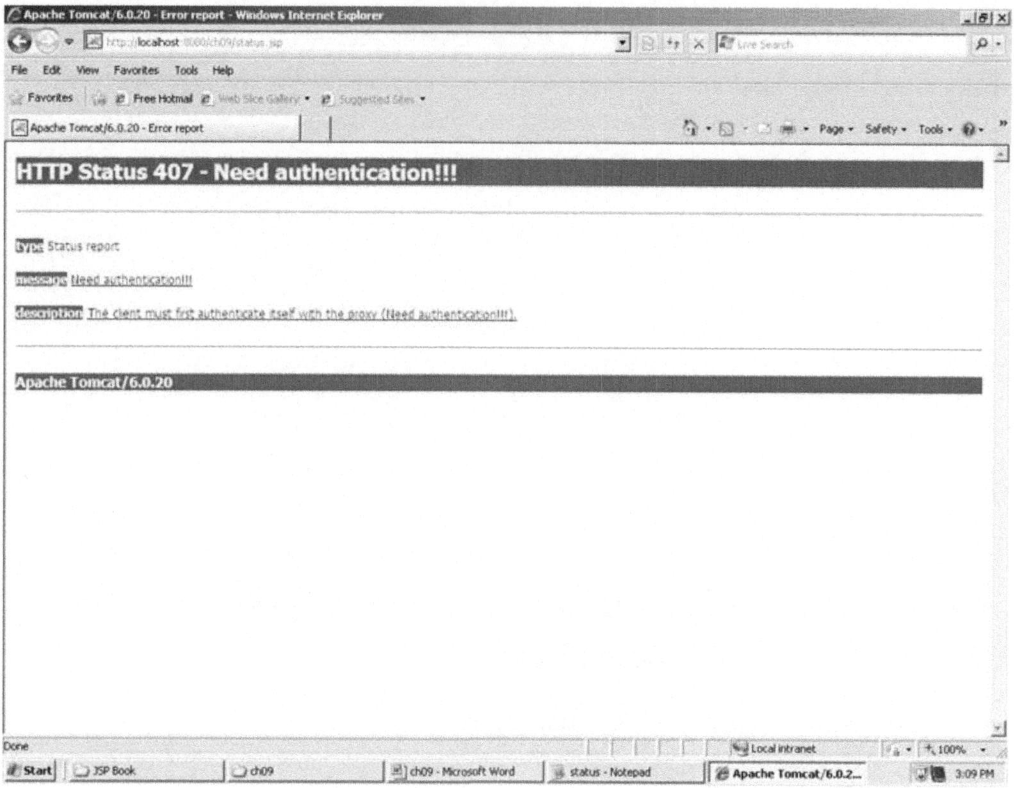

Figure 9.7 Output of status.jsp

This screen (Figure 9.7) display http status information

***Example* 9.6**

GET method example using URL

get.jsp

```
<html>
<head>
<title>using get method to read form data</title>
</head>
```

```
<body>
<center>
<h1>using get method to read form data</h1>
<ul>
<li> <p> <b>first name:</b>
   <%= request.getparameter("first_name")%>
</p> </li>
<li> <p> <b>last  name:</b>
   <%= request.getparameter("last_name")%>
</p> </li>
</ul>
</body>
</html>
```

***Example* 9.7**

GET method example using Form

getform.jsp

```
<html>
<body>
<form action="get.jsp" method="get">
first name: <input type="text" name="first_name">
<br />
last name: <input type="text" name="last_name" />
<input type="submit" value="submit" />
</form>
</body>
</html>
```

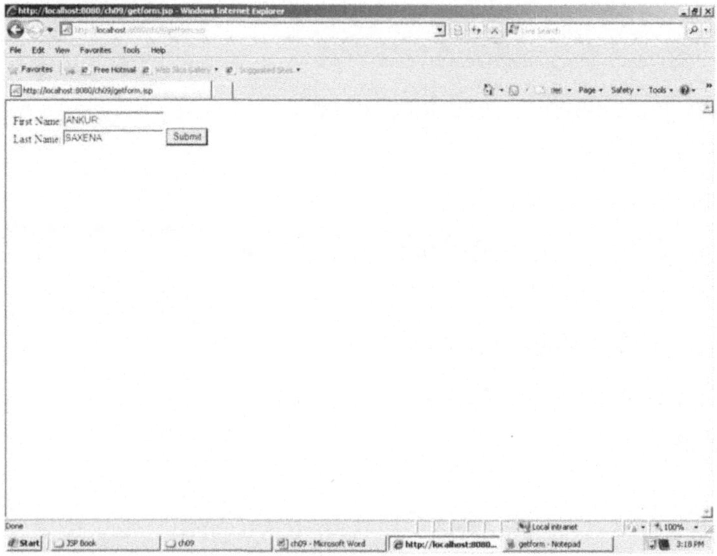

Figure 9.8 Output of getform.jsp

This screen (Figure 9.8) display firstname and lastname now enter firstname and lastname and press submit.

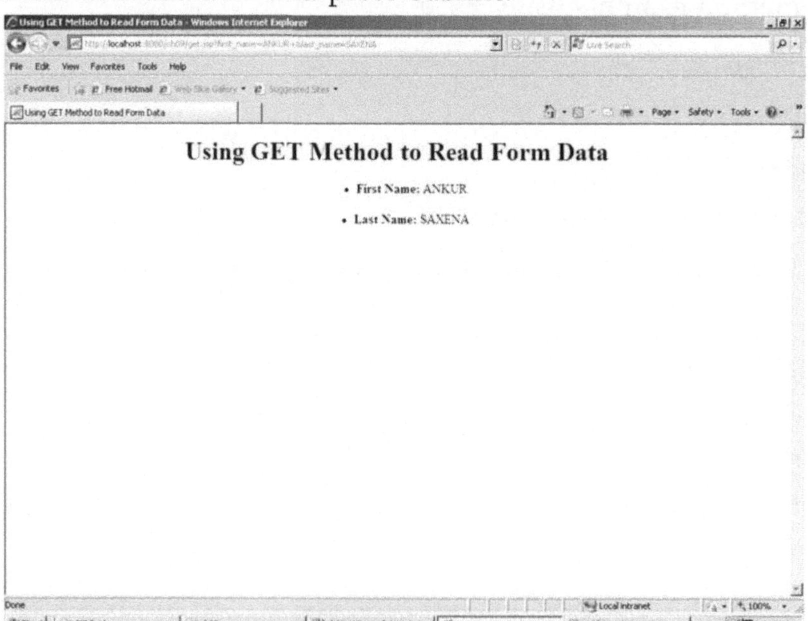

Figure 9.9 Output of get.jsp

This screen (Figure 9.9) display firstname and lastname entered by previous screen with get method.

Example 9.8

POST Method Example Using Form:post.jsp

```
<html>
<head>
<title> post method to read form data</title>
</head>
<body>
<center>
<h1>using get method to read form data</h1>
<ul>
<li> <p> <b>first name:</b>
   <%= request.getparameter("first_name")%>
</p> </li>
<li> <p> <b>last  name:</b>
   <%= request.getparameter("last_name")%>
</p> </li>
</ul>
</body>
</html>
```

Example 9.9

postform.jsp

```
<html>
<body>
<form action="post.jsp" method="post">
first name: <input type="text" name="first_name">
<br />
last name: <input type="text" name="last_name" />
<input type="submit" value="submit" />
</form>
</body>
</html>
```

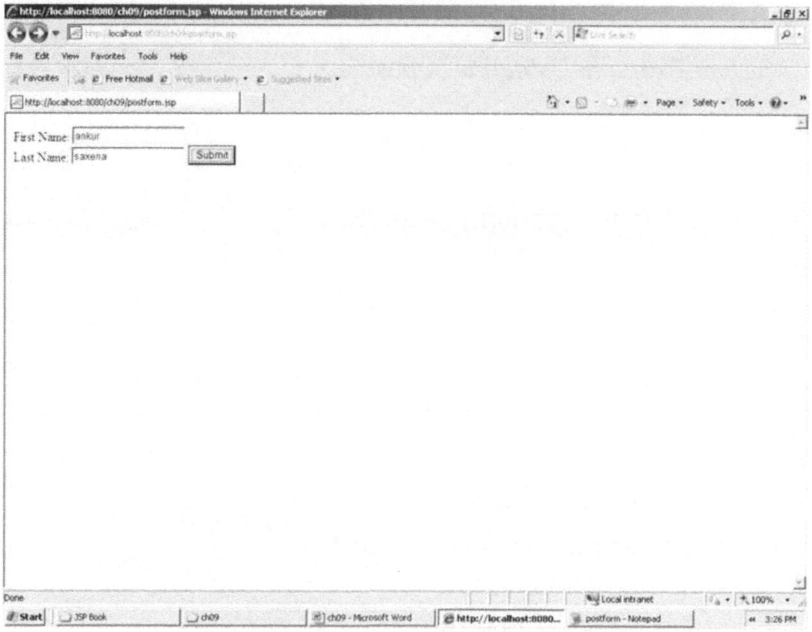

Figure 9.10 Output of postform.jsp

This screen (Figure 9.10) display firstname and lastname. Now enter firstname and lastname and press submit

Figure 9.11 Output of post.jsp

This screen (Figure 9.11) display firstname and lastname entered by previous screen with post method

9.3.4 Out Object

The JspWriter object contains most of the same methods as the java.io.PrintWriter class. However, JspWriter has some additional methods designed to deal with buffering. Unlike the PrintWriter object, JspWriter throws IOExceptions.

The initial JspWriter object is instantiated differently depending on whether the page is buffered or not. Buffering can be easily turned off by using the buffered='false' attribute of the page directive.

The out implicit object is an instance of ajavax.servlet.jsp.JspWriter object and is used to send content in response.

Methods in out object

- print
- println
- newLine
- clear
- clearBuffer
- flush
- is AutoFlush
- getBufferSize
- getRemaining.

print() The print method of out object writes the value to the output without a newline character.

If a JSP program has the following statements :

```
out.print("Welcome!!!");
out.print("Ankur Saxena");
```

println() The println method of out object writes the value to the output with a newline character.

```
out.println("Welcome!!!");
out.println("Ankur");
```

newLine The 'newLine' method of out object is used to write a newline character to the output.

If a JSP program has the following statements

```
out.write("Welcome!!!");
out.newLine();
out.write("Ankur Saxena");
```

clear The clear method of out object is used to clear the output buffer. This method does not write any contents to the client. An exception is thrown by this method if the buffer was flushed.

Syntax `out.clear();`

clearBuffer The 'clearBuffer' method of out object is used to clear the output buffer. This method does not write any contents to the client. The only difference between the 'clear' method of out object and 'clearBuffer' method is

- 'clear' method throws an exception when the buffer is flushed.
- 'clearBuffer' method does not throw an exception when the buffer is flushed.

```
if (out.getBufferSize() != 0)
        out.clearBuffer();
```

flush Two methods of out object, 'clear' and 'clearBuffer' are used to clear the output buffer without writing any contents to the client. The 'flush' method of out object is used to flush the buffer by writing the contents to the client.

```
out.flush();
<jsp:forward page="ank.jsp&">
```

In the above example, no forwarding will take place. This is because of the "out.flush()" method. This method will flush the "JspWriter" buffer. Once the content is written to the browser, a forward cannot be invoked.

isAutoFlush() The "is AutoFlush" method of 'out' object returns a 'true' value if the output buffer is automatically flushed otherwise returns 'false'.

Syntax out.isAutoFlush() ;

getBufferSize The "getBufferSize" method of 'out' object is used to return the size of the buffer. The returned value of the size of the buffer is in bytes. If the output is not buffered, then the "getBufferSize" method returns a 0 byte.

```
if (out.getBufferSize() != 0)
    out.getBufferSize()
```

getRemaining() The "getRemaining" method of out object is used to return the number of empty bytes in the buffer.

Syntax out.getRemaining() ;

Write the value of a session attribute to response.

```
<%
out.write(session.getattribute("username"));
 %>
```

out.jsp

```
<html>
<body>
<%
// this scriptlet declares and initializes "date"
system.out.println( "evaluating date now" );
java.util.date date = new java.util.date();
%>
hello! the time is now
<%
out.println( date );
out.println( "<br>your machine's address is " );
out.println( request.getremotehost());
%>
</body>
</html>
```

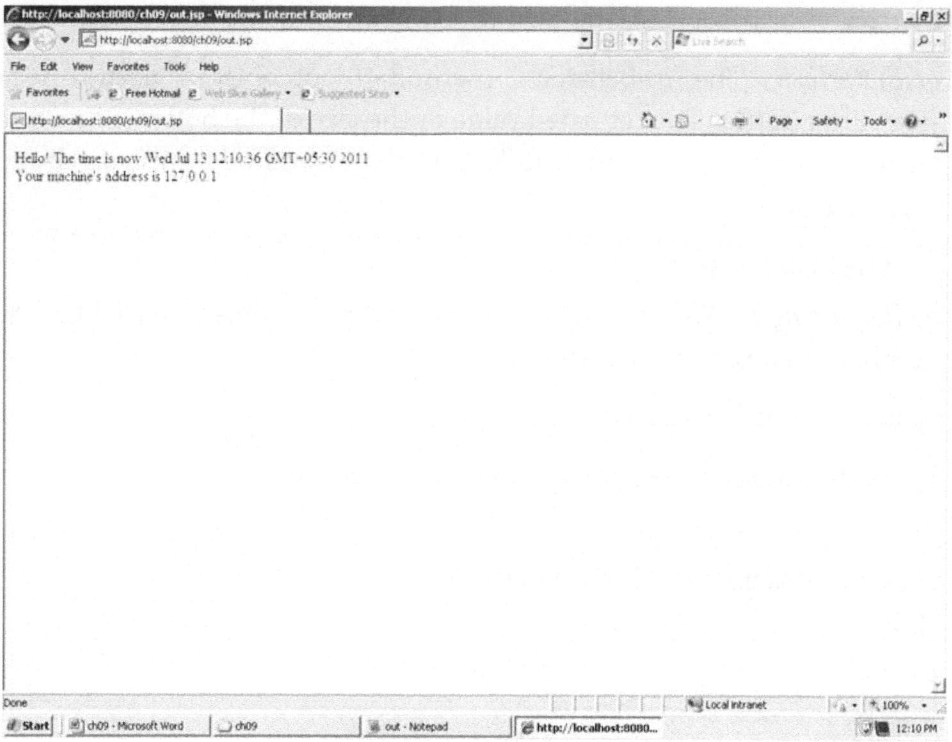

Figure 9.12 Output of out.jsp

This screen (Figure 9.12) display current date and machine address.

Example 9.10

This example display some information using out object

index1.jsp

```
<%@page contentType="text/html" pageEncoding="UTF-8"%>
<%@page import="java.util.List" %>
<%@page import="java.util.ArrayList" %>
<html>
  <head>
     <title>Implicit Out Object</title>
     <style>
        * { font-size: 12px; font-family: verdana }
        td { border: 1px solid #ccc; padding: 3px }
```

```
        th { border: 1px solid #4b8699; padding: 3px;
            background: #4b8699; color: white }
    </style>
  </head>
  <body>
    <h2>out object</h2>
    <%!
        list authors = new arraylist();
    %>
    <%
        if (authors.isempty()) {
          authors.add("ankur saxena");
          authors.add("ankur chaurasia");
          authors.add("ankit");
          authors.add("rahul");
        }
        out.println("<table>");
        out.println("<th>");
        out.println("authors");
        out.println("</th>");
        for (int i = 0; i < authors.size(); i++) {
            out.println("<tr>");
            out.println("<td>");
            out.println(authors.get(i));
            out.println("</td>");
            out.println("</tr>");
        }
        out.println("</table>");
    %>
  </body>
</html>
```

index2.jsp (there is another way, how we can print the table. by mixing html and java code.)

```
<%@page contenttype="text/html" pageencoding="utf-8"%>
<%@page import="java.util.list" %>
<%@page import="java.util.arraylist" %>
<html>
    <head>
        <title>implicit out object</title>
        <style>
            * { font-size: 12px; font-family: verdana }
            td { border: 1px solid #ccc; padding: 3px }
            th { border: 1px solid #4b8699; padding: 3px;
                background: #4b8699; color: white }
        </style>
    </head>
    <body>
        <h2>out object</h2>
        <%!
            list authors = new arraylist();
        %>
        <%
            if (authors.isempty()) {
            authors.add("ankur saxena");
            authors.add("ankur chaurasia");
            authors.add("ankit");
            authors.add("rahul");
            }
        %>
        <table>
        <th>authors</th>
        <% for (int i = 0; i < authors.size(); i++) { %>
```

```
        <tr>
          <td>
            <%= authors.get(i) %>
          </td>
        </tr>
      <%   } %>
      </table>
   </body>
</html>
```

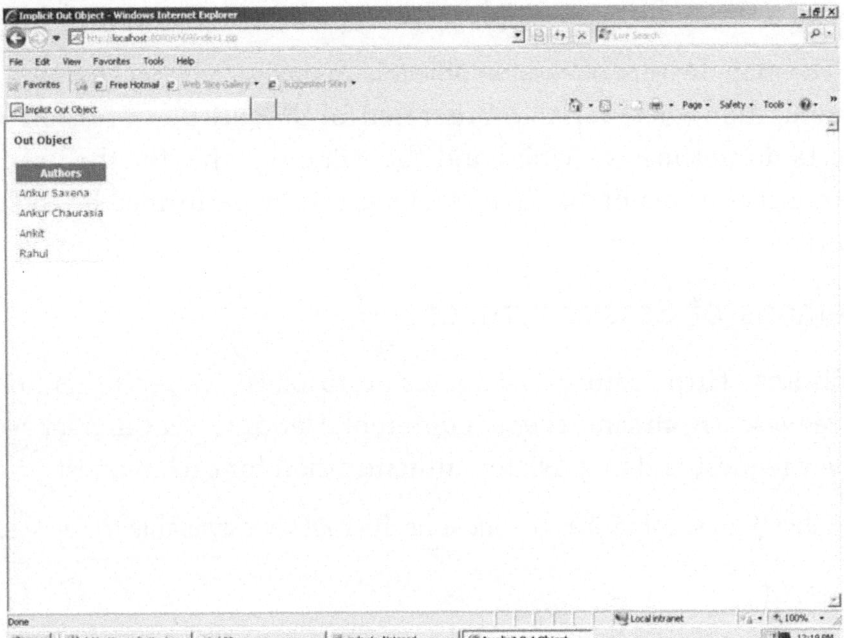

Figure 9.13 Output of index'1'.jsp

This screen (Figure 9.13) display some author name using out object.

9.4 CONTEXTUAL OBJECTS

The implicit objects in this category provide JSP page with access to the context within which it is being processed. The session object, for example, provides the context for the request to which the page is responding. The application object provides the server-side context within which the page is

r u n n i n g .
In contrast, the pageContext object is focused on the context of the JSP page itself, providing programmatic access to all of the other JSP implicit objects which are available to the page, and managing their attributes.

9.4.1 Session Object

Session object represents the data associated with user's session. The request and response, are used to pass information from web browser to server and from server to web browser respectively. While the connection between client and the server is provided by session object. The main use of session object is to maintain states during multiple page requests.

The main feature of session object is to navigate between multiple pages in an application where variables are stored for the entire user session. The session objects do not lose variables and the value remains for the user' s session. The concept of maintenance of sessions can be performed by cookies or URL re-writing.

9.4.2 Methods of Session Object

HttpSession: Http Protocol is stateless protocol. Hence server cannot identify if the request is from same client or different. Hence server can treat every request as new request and may ask for authentication for each request.

Table (9.9) summarises the most useful methods available to the session object.

***Example* 9.11**

Here is a set of pages that put a user's name in the session, and display it elsewhere.

form.jsp

```
<html>
<body>
<form method=post action="session1.jsp">
what's your name? <input type=text name=username size=20>
<p> <input type=submit>
```

Table 9.9 Useful Methods available for a session object

Method	Description
isNew()	The isNew() method of session object returns a true value if the session is new. If the session is not new, then a false value is returned. The session is marked as new if the server has created the session, but the client has not yet acknowledged the session. If a client has not yet chosen the session, i.e., the client switched off the cookie by choice, then the session is considered new. Then the isNew() method returns true value until the client joins the session. Thus, the isNew() method session object returns a Boolean value of true or false.
invalidate()	Discards the session, releasing any objects stored as attributes.
getAttribute(String name)	Retrives the object associated with the named attribute.
getAttributeNames()	The "getAttributeNames" method of session object is used to retrieve all attribute names associated with the current session. The name of each object of the current session is returned. The value returned by this method is an enumeration of object that contains all the unique names stored in the session object.
setAttribute(String, object)	Sets the object to the named attribute, created if it doesnot exist.
removeAttribute(String name)	Removes the object bound with the specified name from this session.
getCreationTime()	Returns the time at which the session was created.
getId	Returns the session ID
getLastAccessedTime	Returns the last time a request associated with the session was received.
getMaxInactiveInterval	Returns the maximum time between requests for which the session will be maintained.

```
</form>
</body>
</html>
```

session1.jsp

```
<%@page language="java" %>
<%
String name = request.getParameter( "username" );
session.setAttribute( "theName", name );
%>
<html>
<head> <title>click to continue</title> </head>
<body>
<a href="sessionresult1.jsp">click to continue</a>
</body>
</html>
```

sessionresult.jsp

```
<%@page language="java" session="true" %>
<html>
<head> <title>dispalying your name</title> </head>
<body>
<font color=red size="10">Hello, <%= session.getAttribute( "theName" ) %> </
    font>
</body>
</html>
```

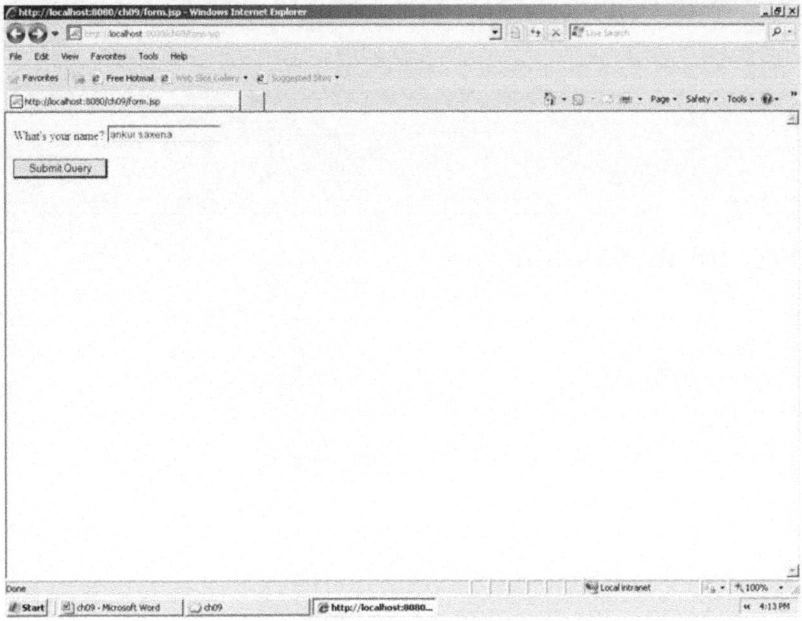

Figure 9.14 Output of form.jsp

This screen (Figure 9.14) shows a text box for name. Now enter your name and press submit query button.

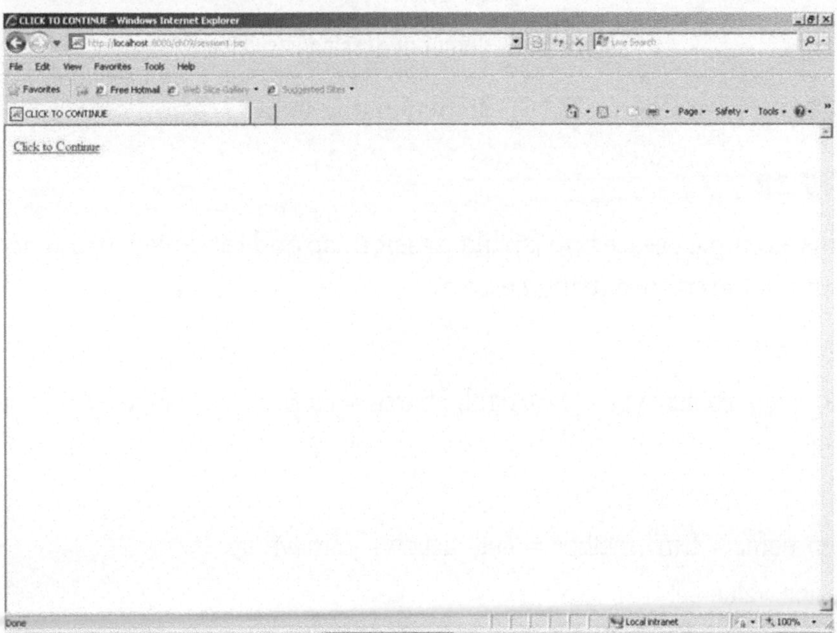

Figure9.15 Output of Session.jsp

This screen (Figure 9.15) show a link to click to continue now press the link.

Figure 9.16 Output of Session result.jsp

This screen (Figure 9.16) display name using session object.

Example 9.12

In this example create two jsp like session.jsp and retrieve.jsp and how to handle set and get attribute using session

session.jsp

```
<%@ page contenttype="text/html; charset=iso-8859-1" language="java" %>
<html>
<body>
<form name="frm" method="get" action="retrieve.jsp">
 session value set
  <input type="text" name="sessionvariable" /> <br>
```

```
<input type="submit" name="submit" value="submit">
</form>
</body>
</html>
```

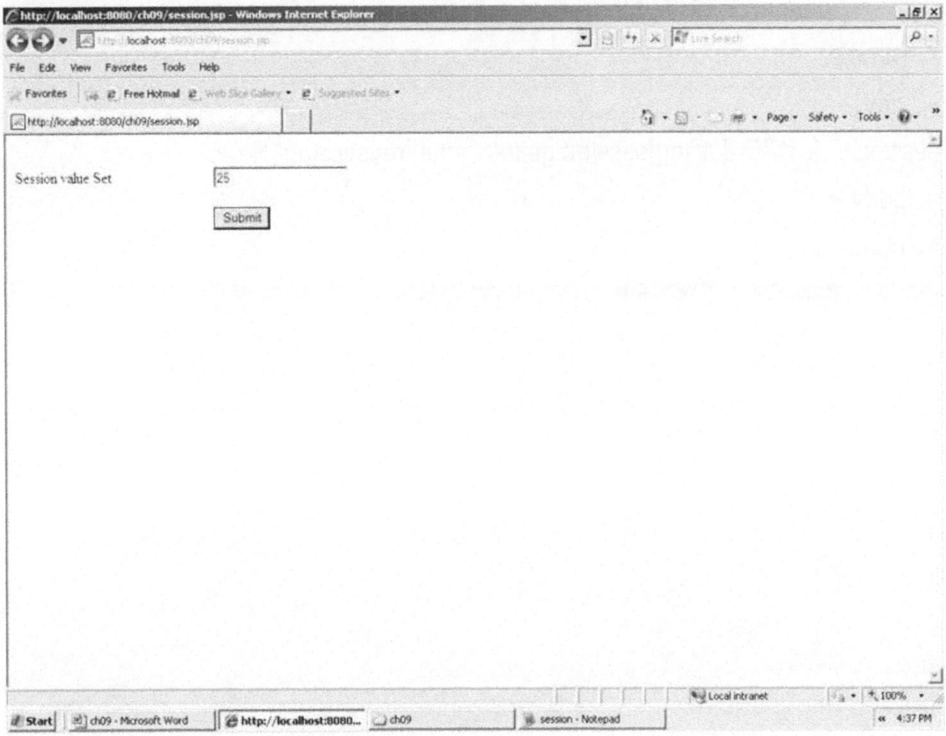

Figure 9.17 Output of Session.jsp

This screen (Figure 9.17) display textbox now enter the value of session and press submit.

***Example* 9.13**

retrieve.jsp

```
<%@ page language="java" import="java.util.*"%>
<%
string sessionset=request.getparameter("sessionvariable");
session.setattribute("mysession", sessionset);
/// string getsessionvalue= (string)session.getattribute("sessionset");
```

```
//this is use for session value in string data
%>
<html>
<head>
<title>cookie example</title>
</head>
<body>
session :    <%=(string)session.getattribute("mysession")%>
</body>
</html>
```

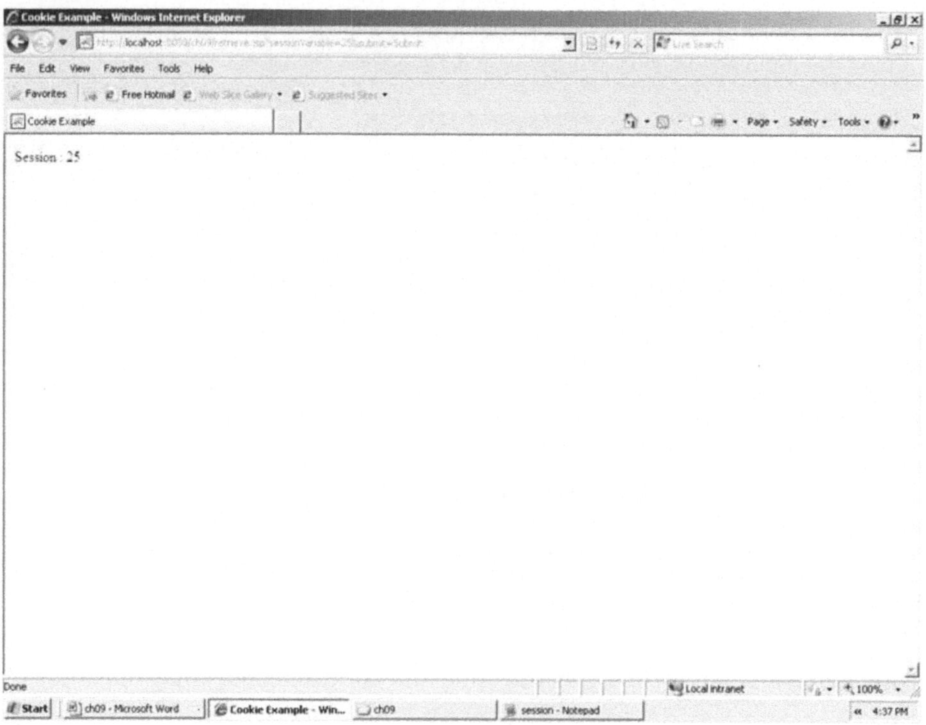

Figure 9.18 Output of retrieve.jsp

This screen (Figure 9.18) display the session value entered by previous screen.

Example 9.14

In this example create two jsp like form2.jsp , session2.jsp and show how to retrieve session object using getValue method.

form2.jsp

```
<html>
<head> <title> log on screen </title> </head>
<body>
<form method="post" action="session2.jsp">
UserName : <input type="text" name="uname">
<br> <br>
Password : <input type="text" name="pass">
<input type="submit"  value="Login">
</form>
</body>
</html>
```

session2.jsp

```
<html>
<head> <title> Retriving Session Objects </title> </head>
<body>
<%
String pass= (String) session.getValue("ID");
String name= (String) session.getValue("NAME");
out.println("Retriving Sessions");
out.println("Name = " + name + "<br> Password = " + pass);
%>
</body>
</html>
```

session3.jsp

This example display session tracking method .

```
<%@ page import="java.io.*,java.util.*" %>
<%
   // Get session creation time.
   Date createTime = new Date(session.getCreationTime());
   // Get last access time of this web page.
```

```
    Date lastAccessTime = new Date(session.getLastAccessedTime());
    String title = "Welcome Back to my website";
    Integer visitCount = new Integer(0);
    String visitCountKey = new String("visitCount");
    String userIDKey = new String("userID");
    String userID = new String("Ankur saxena");
    // Check if this is new comer on your web page.
    if (session.isNew()){
       title = "Welcome to my page";
       session.setAttribute(userIDKey, userID);
       session.setAttribute(visitCountKey,visitCount);
    }
    visitCount = (Integer)session.getAttribute(visitCountKey);
    //visitCount = visitCount + 1;
visitCount++;
    userID = (String)session.getAttribute(userIDKey);
    session.setAttribute(visitCountKey,  visitCount);
%>
<html>
<head>
<title>Session Tracking</title>
</head>
<body>
<center>
<h1>Session Tracking</h1>
</center>
<table border="1" align="center">
<tr bgcolor="#949494">
   <th>Session info</th>
   <th>Value</th>
</tr>
```

```
<tr>
  <td>id</td>
  <td> <% out.print( session.getId()); %> </td>
</tr>
<tr>
  <td>Creation Time</td>
  <td> <% out.print(createTime); %> </td>
</tr>
<tr>
  <td>Time of Last Access</td>
  <td> <% out.print(lastAccessTime); %> </td>
</tr>
<tr>
  <td>User ID</td>
  <td> <% out.print(userID); %> </td>
</tr>
<tr>
  <td>Number of
    visits</td>
  <td> <%
    out.print(visitCount);
    %> </td>
</tr>
</table>
</body>
</html>
```

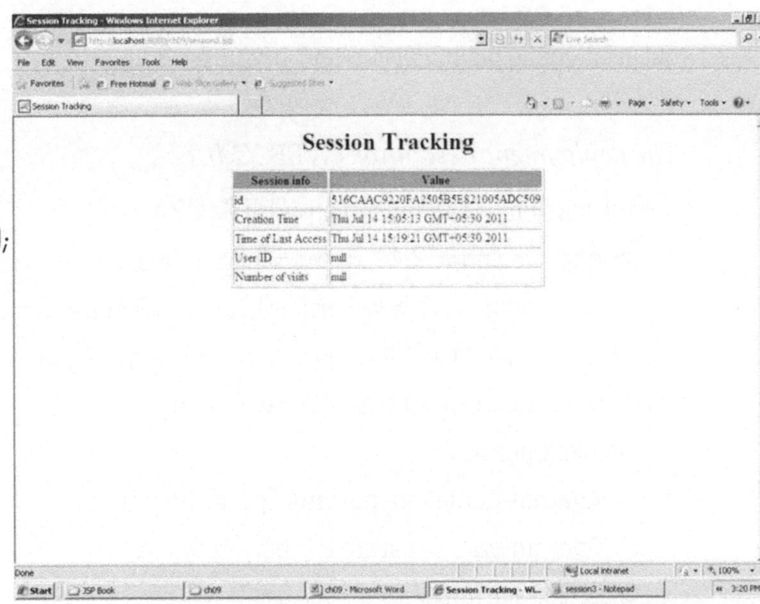

Figure 9.19 Output of Session3.jsp

This screen (Figure 9.19) display Session information and session value.

9.4.3 Application Object

The application object is a direct wrapper around the ServletContext object for the generated Servlet and in reality an instance of a javax.servlet.ServletContext object. This object is a representation of the JSP page through its entire lifecycle. This object is created when the JSP page is initialized and will be removed when the JSP page is removed by the jspDestroy() method. By adding an attribute to application, you can ensure that all JSP files that make up your web application have access to it.

In Apache Tomcat, a web application will be contained in an individual directory under the webapps directory. A typical use of the application object is to get access to the application wide initialisation parameters. Because the application and config objects have similar and slightly confusing uses, the following code shows how they can be used along with a deployment descriptor (WEB.XML) that sets up the appropriate initialisation parameters for them.

index.jsp

```
<%
out.print(config.getInitParameter("myparam2"));
%>
```

The deployment descriptor (WEB.XML)

```
<?xml version="1.0" encoding="UTF-8"?>
<web-app version="2.4" xmlns="http://java.sun.com/xml/ns/j2ee"
 xmlns:xsi="http://www.w3.org/2001/XMLSchema-instance"
 xsi:schemaLocation="http://java.sun.com/xml/ns/j2ee
 http://java.sun.com/xml/ns/j2ee/web-app_2_4.xsd">
   <context-param>
      <param-name>myparam</param-name>
      <param-value>param1</param-value>
   </context-param>
   <servlet>
   <servlet-name>
      MyServlet
```

```
    </servlet-name>
    <jsp-file>
      /index.jsp
    </jsp-file>
    <init-param>
      <param-name>myparam2</param-name>
      <param-value>param2</param-value>
    </init-param>
    </servlet>
    <servlet-mapping>
     <servlet-name>
        MyServlet
     </servlet-name>
     <url-pattern>
        /index.jsp
     </url-pattern>
    </servlet-mapping>
</web-app>
```

Methods of implicit object application

- getAttribute(String name)
- getAttributeNames
- setAttribute(String objName, Object object)
- removeAttribute(String objName)
- getMajorVersion()
- getMinorVersion()
- getServerInfo()
- getInitParameter(String name)
- getInitParameterNames
- getResourceAsStream(Path)
- log(Message)

getAttribute(String name) This method returns the 'object' stored in specified attribute name. If the specified parameter does not exist, it will return 'null' value.

Example application.getAttribute("Ankur Saxena");

getAttributeNames() This method returns all the attributes available within the application. The return type is Enumeration.

Example Enumerationankur;

ankur = application.getAttributeNames();

setAttribute(String objName, Object object) This method stores the 'object' in the object name provided(String objname) in the application.

Example application.setAttribute("Variable", hello);

removeAttribute(String objName) This method is used to remove the specified attribute (String objName) from the application.

Example application.setAttribute("password",password);

application.removeAttribute("password");

getMajorVersion() This method is used to return the major version of the Servlet API for the JSP Container.

Example out.println("Major Version:"+application.getMajorVersion());

Output Major Version:2

The above statement gives 2 as the major version of the Servlet API in use for the Application object.

getMinorVersion() This method is used to return the minor version of the Servlet API for the JSP Container.

Example out.println("MinorVersion:"+application.getMinorVersion());

Output Minor Version:1

The above statement gives 1 as the minor version of the Servlet API in use for the Application object.

getServerInfo() The method 'getServerInfo' of Application object is used to return the name and version number of the JRun servlet engine. Information about the JSP container, such as, the name and product version, are returned by the method getServerInfo of Application object.

Example out.println("Server Information:"+application.getServerInfo());

getInitParameter(String name) The method 'getInitParameter' of application object is used to return the value of an initialization parameter. If the parameter does not exist, then null value is returned.

Example String ank = application.getInitParameter("eURL");

In the above example, the value of initialization parameter "eURL" is retrieved and stored in string "ank".

getInitParameterNames() The method 'getInitParameterNames' of application object is used to return the name of all initialization parameter of the application. The returned value is an enumeration.

Example Enumeration e; *e=application.getInitParameterNames();*

getResourceAsStream(Path) The method 'getResourceAsStream' of application object is used to translate the resource URL mentioned as parameter in the method , into an input stream to read.

Example InputStream stream = application.getResourceAsStream("/ankur.txt")

log(Message) The method log of application object is used to write a text string to the JSP Container' s default log file. General syntax of log method of Application object is as follows.

```
application.log(Message);
```

Assign value of a context init parameter to a variable.

```
<%
 String param1 = application.getInitParameter("param1");
 %>
```

The servlet context obtained from the servlet configuration object (as in the call this.getServletConfig().getContext()) has an application scope.

```
<%
      javax.servlet.RequestDispatcher rd;
      rd = application.getRequestDispatcher("/index.jsp");
      rd.forward(request, response);
%>
```

***Example* 9.15**

This example show recourse path using application object.

application.jsp

```
<%@page contentType="text/html" pageEncoding="UTF-8"%>
<html>
  <head>
     <title>application object</title>
     <style>
        * { font-size: 12px; font-family: Verdana }
     </style>
  </head>
  <body>
     <h2>application object</h2>
     <p>
        Resource paths:
     </p>
     <%= application.getResourcePaths("/")  %> <br>
     <%= application.getResourcePaths("/WEB-INF")  %><br>
     <%= application.getResourcePaths("/META-INF")  %>
     <p>
        Server:
     </p>
     <%= application.getServerInfo()  %>
     <p>
        Context path:
```

```
        </p>
        <%= application.getContextPath() %>
    </body>
<html>
```

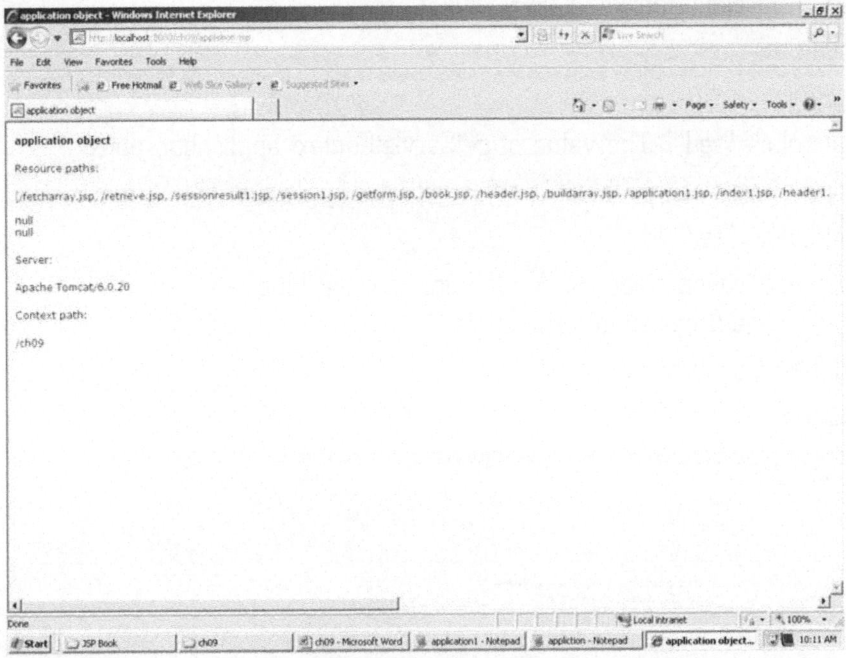

Figure 9.20 Output of application.jsp

This screen (Figure 9.20) display resource path of the url.

Example 9.16

This example show context information of the page.

application1.jsp

```
<%@ page language="java" import="java.sql.*" %>
<%
String parameterValue="This is application object";
getServletContext().setAttribute("ApplicationValue","This is my Name");
StringgetValueApplcation=(String)getServletContext().getAttribute("ApplicationValue");
application.setAttribute("ApplcationVariable",parameterValue);
```

```
String getApplicationVariable=(String)application.getAttribute("ApplcationVariable");
%>
<html>
<head>
<title>Application object in JSP</title>
</head>
<body>
<font color="red">The value of getServletContext application object :
</font> <font color="blue"> <%=getValueApplcation%> </font> <br>
<font color="red">
This is application object :</font> <font color="blue">
    <%=getApplicationVariable%>
</font> </body>
</html>
```

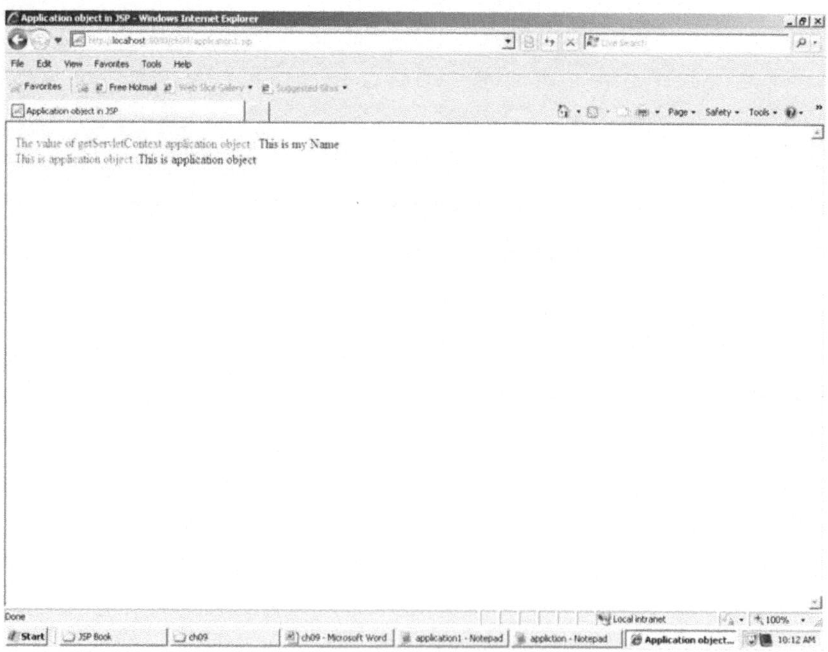

Figure 9.21 OUtput if applicationl.jsp

This screen (Figure 9.21) display context result of the page.

Example 9.17

This example show hit counter integer result using application object.

application2.jsp

```
<%@ page import="java.io.*,java.util.*" %>
<html>
<head>
<title>Applcation object in JSP</title>
</head>
<body>
<%
   Integer hitsCount =
     (Integer)application.getAttribute("hitCounter");
   if( hitsCount ==null || hitsCount == 0 ){
     /* First visit */
     out.println("Welcome to my page");
     hitsCount = 1;
   }else{
     /* return visit */
     out.println("Welcome back to my page");
     hitsCount += 1;
   }
   application.setAttribute("hitCounter", hitsCount);
%>
<center>
<p>Total number of visits: <%= hitsCount%></p>
</center>
</body>
</html>
```

Figure 9.22 Output of application2.jsp

This screen (Figure 9.22) display the total number of visits in current page.

9.4.4 Page Context Object

A PageContext instance provides access to all the namespaces associated with a JSP page, provides access to several page attributes, as well as a layer above the implementation details.

The PageContext provides a number of facilities to the page/component and page implementor, including

- a single API to manage the various scoped namespaces
- a number of convenience API's to access various public objects
- a mechanism to obtain the JspWriter for output
- a mechanism to manage session usage by the page
- a mechanism to expose page directive attributes to the scripting environment
- mechanisms to forward or include the current request to other active components in the application

- a mechanism to handle "errorpage" exception processing

It can be seen as a convenience object for tasks such as transferring requests to other web resources. Much of its functionality appears to be for the convenience of generating the servlet from JSP, thus the following example of a servlet generated from the Tomcat container shows how PageContext is used to populate other implicit objects.

```
application = pageContext.getServletContext();
config = pageContext.getServletConfig();
session = pageContext.getSession();
out = pageContext.getOut();
```

A typical use of the pageContext is to to include another resource or forward to another resource. Thus the following would forward from the current page to menu.jsp

```
<%
pageContext.forward("menu.jsp");
%>
```

pageContext instance that contains all data's information, which is associated with JSP page. This helps to access attribute and other field's object set in JSP page. pageContext is in javax.servlet.jsp.PageContext class. pageContext is to manage various scope of attribute in shared information object.

Table 9.10 summarises the method of pageContext.

Set an attibute in pagecontext

```
<%
pagecontext.setAttribute("name", "servletworld");
%>
```

Access value of a pagecontext attribute and output to response.

```
<%= pagecontext.getAttribute("name") %>
```

One of the important methods is removeAttribute, which accepts either one or two arguments. For example, pageContext.removeAttribute ("attrName") removes the attribute from all scopes, while the following code only removes it from the page scope.

Table 9.10 Methods of Page Context

Method	Return	Description
pageContext.findAttribute(String Name)	Object	Search for named attribute in JSP page Request, session, application scope .
pageContext.forward(String relativeURLPath)	void	This method is used to redirect or forward th current ServletRequest or ServletResponse to another active component .
pageContext.getAttribute(String Name)	Object	Returns the object associated with name in the page scope.
pageContext.getAttributesScope(String Name)	int	Get the scope where a given attribute is defined.
pageContext.getErrorData()	ErrorData	Provide access to error information.
pageContext.getException()	Exception	Get current value of the exception Object.
pageContext.getOut()	JspWriter	Get current value of out object.
pageContext.getRequest()	ServletRequest	Get current value of request object.
pageContext.getResponse()	ServletResponse	Get current value of response object.
pageContext.getServletConfig()	ServletConfig	Get any initialization parameters and startup configuration for this servlet.
pageContext.getServletContext()	ServletContext	Gets the context from the servlet's ServletConfig object.
pageContext.getSession()	HttpSession	Return the current value of session object.

```
pageContext.removeAttribute("attrName",page_scope);
```

***Example* 9.18**

This example show how to save and retrieve data using page context object

pagecontext.jsp

```
<b> <font color="purple">
<%
// Check if attribute has been set
Object o = pageContext.getAttribute("ankur.amity.name1", pagecontext.page_scope);
if (o == null) {
out.println("'ankur.amity.name1' is not 'null'");
}
// Save data
pageContext.setAttribute("ankur.amity.name0", "value0")  // page_scope is the default
pagecontext.setattribute("ankur.amity.name1", "value1", pagecontext.page_scope);
pagecontext.setattribute("ankur.amity.name2", "value2", pagecontext.request_scope);
pagecontext.setattribute("ankur.amity.name3", "value3", pagecontext.session_scope);
pagecontext.setattribute("ankur.amity.name4", "value4",
      pagecontext.application_scope);
out.println("<br>");
out.println("<br>");
out.println("showing values of attribute :-");
%>
<br>
<%-- show the values --%>
<%= pagecontext.getattribute("ankur.amity.name0") %> <%-- page_scope --%>
<br>
<%= pagecontext.getattribute("ankur.amity.name1", pagecontext.page_scope) %>
<br>
<%= pagecontext.getattribute("ankur.amity.name2", pagecontext.request_scope) %>
<br>
<%= pagecontext.getattribute("ankur.amity.name3", pagecontext.session_scope) %>
```

```
<br>
<%= pagecontext.getattribute("ankur.amity.name4", pagecontext.application_scope)
    %>
</font> </b>
```

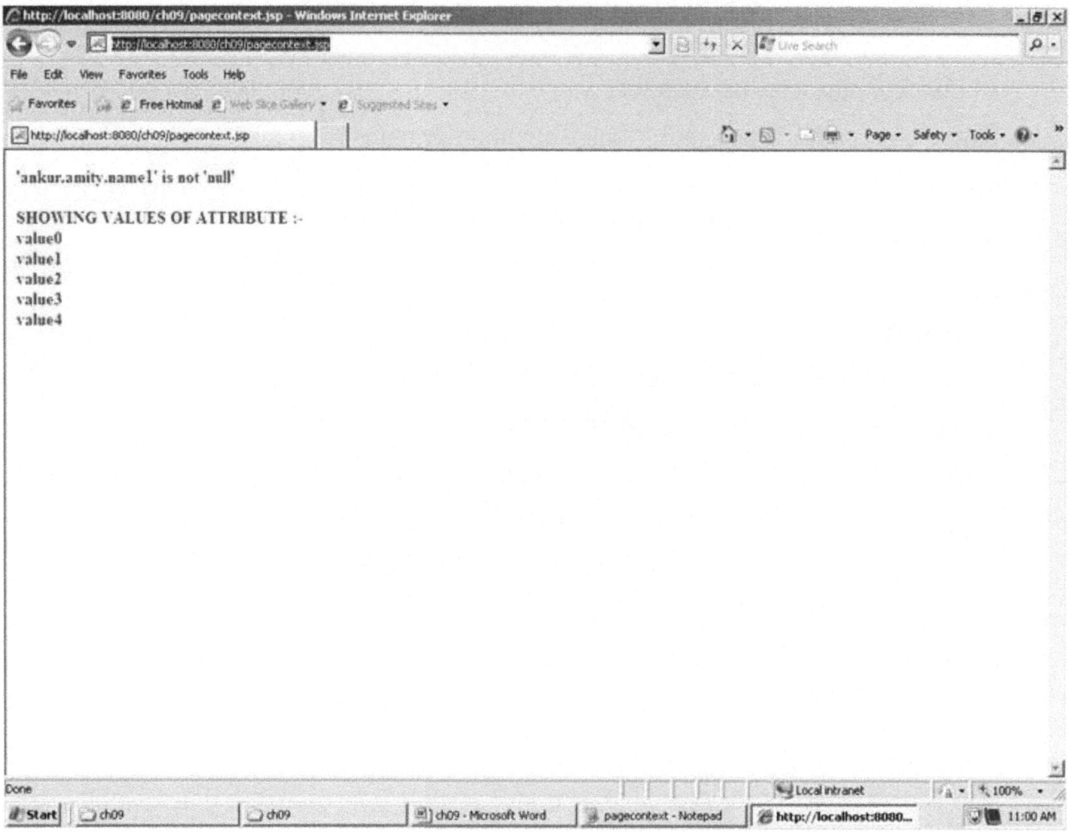

Figure 9.23 Output of pagecontext.jsp

This screen (Figure 9.23) display the values of saved data.

Difference between page and pagecontext

page The page and pageContext are implicit JSP Objects. The page object represents the generated servlet instance itself, i.e., it is same as the "this" keyword used in a Java file. As a result, you do not typically know who the super class is, and consequently do not normally make use of this object or its methods. page is of class java.lang.Object and it refers to instance of generated servlet. It is declared as,

Object page this;

// this refers to the instance of this servlet

page cannot be used to directly call the servlet methods.

1. < page.getServletInfo() >

   Error because page is java.lang.Object type
2. < ((servlet)page).getServletInfo() > <----OK:typecast
3. < this.getServletInfo() > <-------OK

pageContext The pageContext object represents the environment for the page, containing useful information like page attributes, access to the request, response and session objects, as well as the JspWriter referenced by out. This object also has methods for including another URL's contents, and for forwarding or redirecting to another URL.

pageContext is of type 'javax.servlet.jsp.PageContext.'

pageContext class is an abstract class.

The key things it do as are

1. provide convenient methods to get and set attributes in diff scopes.
2. provide convenient methods for transfering requests to other resources in the web application.

   void include(String relativeURL)& void forward(String relativeURL)

   Ex pageContext.forward(ankit.jsp);
3. store the references to implicit objects.

Table 9.11 Summarises the difference between page and pageContext. page vs pageContext

Implicit object	Class or interface	Purpose, function, or uses
page	java.lang.Object	Refers to the generated Servlet class. Since page refers to the generated class and the class implements the Servlet interface, it is valid cast if to Servlet. And the page variable could also be cast to JspPage or HttpJspPage since these two interfaces are derived from the Servlet interface and are implemented by the generated servlet class. For example, <%= ((Servlet)page).getServletInfo () %>

pageContext	javax.servlet.
	jsp.PageContext
		Used for storing and retrieving page-related information and sharing objects within the same translation unit and same request.
		Also used as a convenience class that maintains a table of all the other implicit objects. For example,
		public void _jspService (
		    HttpServletRequest request,
		    HttpServletResponse response) throws java.io.IOException,
		ServletException {
		...
		  try {
		   ...
		  application = pageContext.getServletContext();
		  config = pageContext.getServletConfig();
		 session = pageContext.getSession();
		  out = pageContext.getOut();
		   ...
		  } catch (Throwable t) {
		   ...
		  } finally {
		   ...
		  }
		}

9.5 ERROR HANDLING

This final category of JSP implicit objects has only one member, the exception object. This object is provided for the purpose of error handling within JSP.

9.5.1 Exception Object

The exception object represents all errors and exceptions. The exception implicit object is of type java.langThrowable. You can access the exception object on a page that you declare to be an error page using the isErrorPage attribute of the page directive. Note that the exception object is created only if the JSP uses the page directive to is ErrorPage set to true. When a JSP generates an error and forwards that error to the error page, it sets the JSP exception object of the error page to the generated error.

```
<%@ page isErrorPage='true' %>
```

The exception object is an instance of a subclass of Throwable (e.g., java.lang. NullPointerException) and is only available in error pages. Table 9.12 lists of the important medthods available in the Throwable class.

***Example* 9.19**

This example display how to trace exception in jsp.

trace.jsp

```
<%@ page isErrorPage="true" %>
<html>
<head>
  <title>Example - Implicit Object "exception"</title>
</head>
<body>
<pre>
  Message: <%= exception.getMessage() %>
  Stack Trace:
  <%
  StackTraceElement[] stackTrace = exception.getStackTrace() ;
```

```
for ( int i = 0 ; i < stackTrace.length ; ++i ) {
%>
   <%= stackTrace[i] %>
```

Table 9.12 List of Methods available in throwable class

SN	Methods with Description
1	public String getMessage() Returns a detailed message about the exception that has occurred. This message is initialized in the Throwable constructor.
2	public Throwable getCause() Returns the cause of the exception as represented by a Throwable object.
3	public String toString() Returns the name of the class concatenated with the result of getMessage().
4	public void printStackTrace() Prints the result of toString() along with the stack trace to System.err, the error output stream.
5	public StackTraceElement [] getStackTrace() Returns an array containing each element on the stack trace. The element at index 0 represents the top of the call stack, and the last element in the array represents the method at the bottom of the call stack.
6	public Throwable fillInStackTrace() Fills the stack trace of this Throwable object with the current stack trace, adding to any previous information in the stack trace.

```
 <%
  }
 %>
</pre>
</body>
</html>
```

Example 9.20

Exception Implicit object in JSP

exception.jsp

```
<%@ page  language="java" errorPage="error.jsp" %>
<html>
<head>
<title>Implicit Exception object</title>
</head>
<body>
<%
 String a=null;
 int b=Integer.parseInt(a);
%>
</body>
</html>
```

error.jsp

```
<%@ page  language="java" import="java.io.*" isErrorPage="true"  %>
<html>
<head>
<title>Implicit Exception Error page</title>
</head>
<body>
The Exception is <strong> <%= exception.toString() %> </strong> <br>
```

```
Message : <strong> <%=exception.getMessage()%></strong> <br>
</body>
</html>
```

Example 9.21

The following code demonstrates the use of the exception implicit object. The first page uses the errorPage directive to set up the JSP page to use when an error occurs, and the second page called ErrorPage.jsp uses the isErrorPage directive to set itself up to catch the error.

Exception1.jsp

```
<%@page errorPage="ErrorPage.jsp" %>
<%
  int i=10;
  /*Divide by zero, generates an error */
  out.print(i/0);
%>
```

ErrorPage.jsp

```
<%@ page isErrorPage="true" %>
<%
out.print("<h1> Here is the error message </h1>");
out.print(exception.getMessage());
%>
```

The following tags would re-direct all HTTP 404 errors to the /errorpage.jsp page.

```
<error-page>
   <error-code>404</error-code>
   <location>/errorPage.jsp</location>
</error-page>
```

Example 9.22

main.jsp..

```
<%@ page errorpage="showerror.jsp" %>
<html>
<head>
  <title>error handling example</title>
</head>
<body>
<%
  // throw an exception to invoke the error page
  int x = 1;
  if (x == 1)
  {
     throw new runtimeexception("error condition!!!");
  }
%>
</body>
</html>
```

Now you would have to write one Error Handling JSP ShowError.jsp, which is given below. Notice that the error-handling page includes the directive <%@ page isErrorPage="true" %>. This directive causes the JSP compiler to generate the exception instance variable.

```
<%@ page iserrorpage="true" %>
<html>
<head>
<title>show error page</title>
</head>
<body>
<h1>opps that's a huge error</h1>
<p>sorry, an error occurred.</p>
<p>here is the exception stack trace: </p>
<pre>
<% exception.printstacktrace(response.getwriter()); %>
```

```
</pre>
</body>
</html>
```

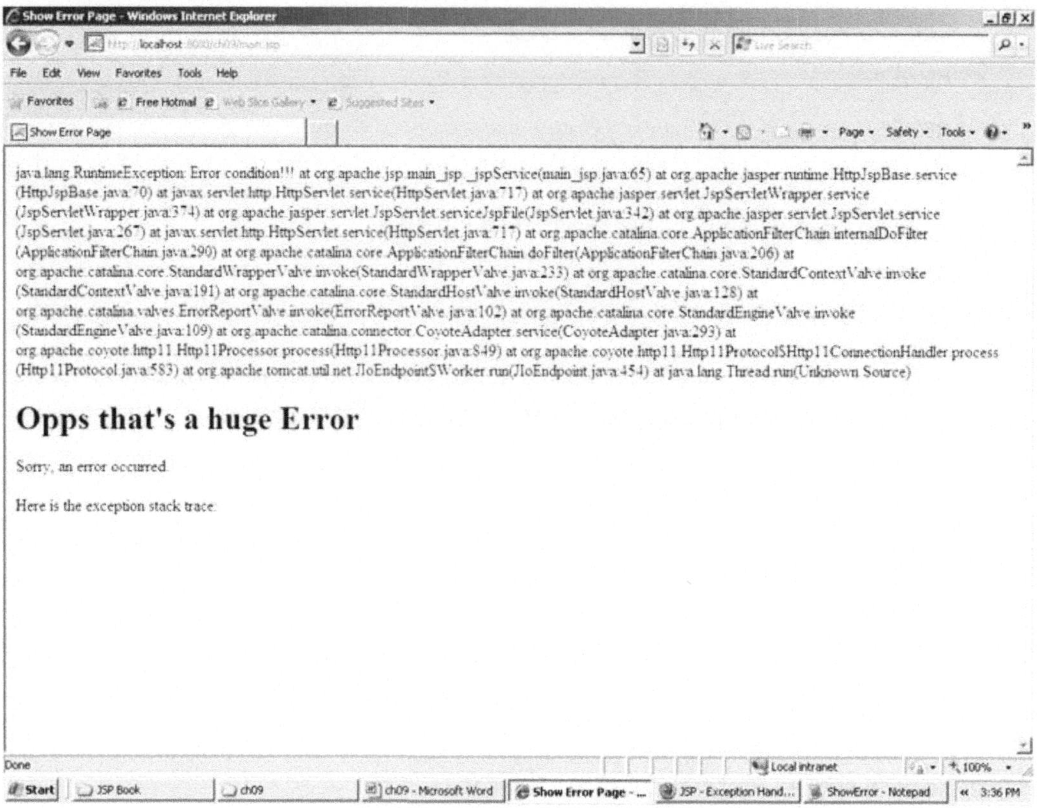

Figure 9.24 Output of main.jsp

This screen (Figure 9.24) display the exception results

Example 9.23

In this example show how to catch a exception in jsp a page.

exception2.jsp

```
<html>
<head>
  <title>try...catch example</title>
</head>
```

```
<body>
<%
  try{
     int i = 1;
     i = i / 0;
     out.println("the answer is " + i);
  }
  catch (exception e){
     out.println("an exception occurred: " + e.getmessage());
  }
%>
</body>
</html>
```

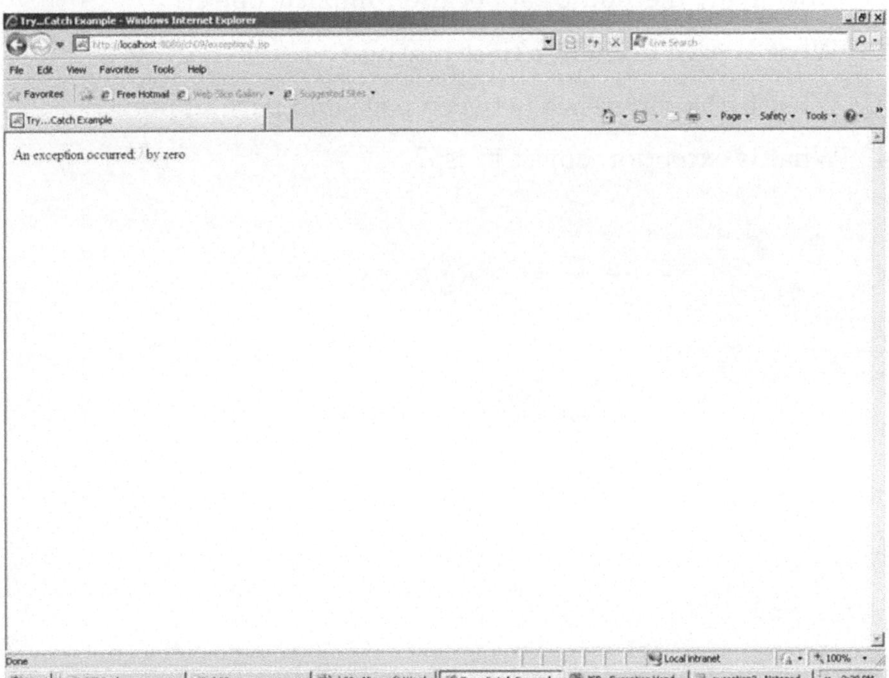

Figure 9.25 Output of exception2.jsp

This screen (Figure 9.25) display the result of occurred exception in jsp page.

REVIEW QUESTIONS

1. What are implicit objects in jsp?
2. What is the method for storing and retrieving attribute values?
3. What are servlet related objects in jsp?
4. What is config object?
5. What is the difference between request and response implicit objects?
6. What is the difference between get and post method?
7. What is `response.sendRedirect()` method?
8. Explain all method of out object?
9. What are contextual objects?
10. What are session objects?
11. How many method of application implicit object?
12. What is page Context object in jsp?
13. What is the difference between page and page Context?
14. What is exception object in jsp?

10

JAVA BEANS WITH JSP

10.1 JAVABEAN

A JavaBean can be defined as a reusable software component. What this means is that we can write a JavaBean that can then be used in a variety of other Java-based softwares such as applications, Servlets or JSP pages. In this way we can define our business logic within a JavaBean and then consistently use that logic in separate applications.

10.1.1 Bean Containers

A bean container is an application, environment or programming language that allow developers to call up Beans. Configure them, and access their information and behaviour. Application that use beans are composed of purely java code, but bean containers allow developers to work with it at a higher conceptual level. The bean container defines its own ways of presenting and interacting with the bean and writes the resulting java code itself. JSP container allow developers to create web-based java applications without needing to write java code. In JSP, we interact with beans through a collection of tags that we can embed inside our HTML.

10.1.2 Benefits of JavaBeans

We now have three ways of writing code to be used by a JSP. They are:

1. Place the code at the start of a JSP in a declaration,
2. Use an include statement to reference another file which contains the code and

3. Now package the code in a JavaBean

Which method do we use and when?

As always, it depends on your circumstances. Writing the code within a JSP page is certainly the most straightforward, but it does limit the amount of code reuse. Using include statements does lend itself to a certain amount of code reuse and certainly it is the only way to allow reuse of chunks of HTML.

By using JavaBeans you can fully separate the business logic from the generation of the display. This is an important philosophy that leads to better structured and more maintainable systems. In our case, you would use the Java Server Pages to dynamically generate display and also to handle the user interaction. The JavaBean would take over when you need to perform some complex data processing or when you need to access databases or the file system.

The other advantage of using JavaBeans is that the business logic can be used by more than one application. For example, both a client-based Java application and a JSP page can access the same JavaBean thus guaranteeing the same functionality.

A final point to note is that by using JavaBeans you can split your development team into Java experts and HTML experts. The Java experts would write and develop the JavaBeans and the HTML experts would concentrate on the design of the web application.

Example 10.1

JavaBean

```
packagecom.bo.jspBean;
publicclassemployeeBean
{
privateStringfirstName="";
 privateStringlastName="";
//FirstNameproperty
 publicvoidsetFirstName(String name)
 {
```

```
 firstName=name;
 }
publicStringgetFirstName()
 {
 returnfirstName;
 }
//LastNameProperty
 publicvoidsetLastName(String name)
 {
 lastName=name;
 }
publicStringgetLastName()
{
 returnlastName;
  }
//FullNameProperty-Read Only
 publicStringgetFullName()
{
returnfirstName+""+lastName;
}
}
```

10.1.3 Compiling a JavaBean

This section gives an overview of the steps involved.

Step 1 **Compile the class files** The installation of the JDK included several utilities that we will use for our JavaBeans. The first utility, javac — the java compiler, is used to compile class files from the java source code files using the following syntax,

```
javac-verbose*.java
```

This option will compile all the ".java" files in the current directory. The "verbose" option is used to generate compilation information. There are other options available and are detailed in the JDK documentation.

***Step* 2 Build the JAR Archive File** Next we run the jar utility that will build the JAR file. It takes the following format,

```
jarcvmfmeta-inf\manifest.mfjarName.jarpackageName\*.class
```

This should be entered as a single line and jarName is the name of the JAR file we are building and packageName is the name of the folder in which our class files reside (again it is also the name of the package). The JAR file is usually given the same name as the package.

Note if the package name is structured like com\ankur then the "." need to be replaced with "\" when writing the jar command. For example,

```
jarcvmfmeta-inf\manifest.mfjarName.jarcom\ankur\*.class
```

The cvmf clause are options available to the JAR utility. The c instructs the utility to create a JAR file, v forces verbose output of the compilation, m indicates that we are providing a manifest file and f indicates that we are providing the JAR file name.

***Step* 3 Deploying the JavaBean** Once we have created our JavaBean, we then need to place the file in a suitable location on our hard drive, so that it can be accessed by the Application Server.

If you are using the JavaBean unpackaged then place the file under the classes folder of the WEB-INF folder. If you are using a package name such as com\ankur then you need to create this folder structure under "WEB-INF/classes"suchas"WEB-INF/classes/com/ankur/mybean.class".

If you have packaged the JavaBean in a jar file then place the file in the "WEB-INF/lib" folder.

You may need to restart your Application Server before the JavaBean gets recognized.

10.2 JSP ACTIONS FOR JAVABEANS

To use a JavaBean in our JSP page we make use of three JSP actions that we saw earlier. They are,

- <jsp:useBean/> find or instantiate a JavaBean
- <jsp:setProperty/> sets the property of a JavaBean

- <jsp:getProperty/> gets the value of a JavaBean property

The first of these is used to specify which JavaBean we want to use in our page. The next two actions, as their names suggest, are used to set and get values of properties of the JavaBean.

10.2.1 jsp:useBean

The <jsp:useBean> element locates or instantiates a JavaBeans component. <jsp:useBean> first attempts to locate an instance of the bean. If the bean does not exist, <jsp:useBean> instantiates it from a class or serialized template.

To locate or instantiate the bean, <jsp:useBean> takes the following steps, in this order.

1. Attempts to locate a bean with the scope and name you specify.
2. Defines an object reference variable with the name you specify.
3. If it finds the bean, it stores a reference to it in the variable. If you specified type, it gives the bean that type.
4. If it does not find the bean, <jsp:useBean> instantiates it from the class you specify, storing a reference to it in the new variable. If the class name represents a serialized template, the bean is instantiatedbyjava.beans.Beans.instantiate.
5. If jsp:useBean has instantiated (rather than located) the bean, and if it has body tags or elements (between <jsp:useBean> and </jsp:useBean>), then it executes the body tags.

The body of a jsp:useBean element often contains a jsp:setProperty element that sets property values in the bean. As described in Step 5, the body tags are only processed if jsp:useBean instantiates the bean. If the bean already exists and jsp:useBean locates it, the body tags have no effect.

You can use a jsp:useBean element to locate or instantiate a JavaBeans component, but not an enterprise bean. To create enterprise beans, you can write a jsp:useBean element that calls a bean that in turn calls the enterprise bean, or you can write a custom tag that calls an enterprise bean directly.

JSP syntax The JSP syntax of jsp:useBean is given below.

```
<jsp:useBeanid="beanInstanceName"
 scope="page|request|session|application"
{
 class="package.class"[type="package.class"]|
 beanName="{package.class |'${'Expression'}'|
 <%= expression %>}" type="package.class"|
 type="package.class"
}
{/>|>otherelements</jsp:useBean>}
```

XML syntax The XML syntax of jsp:useBean is given below

```
<jsp:useBean id="beanInstanceName"
scope="page|request|session|application"
{
 class="package.class"[type="package.class"]|
 beanName="{package.class|'${'Expression'}'|
 %= expression %}"type="package.class"|
 type="package.class"
}
{/>|>otherelements</jsp:useBean>}
>
```

Attributes and usage The attributes of jsp:useBean are listed below

id="beanInstanceName" A variable that identifies the bean in the scope you specify. You can use the variable name in expressions or scriptlets in the JSP page.

The name is case-sensitive and must conform to the naming conventions of the scripting language used in the JSP page.

scope="page|request|session|application"

page scope Any object whose scope is the page will disappear as soon as the current page finishes generating. The object with a *page scope* may be modified as often as desired within the particular page but the changes are lost as soon as the page exits.

By default all beans have *page scope*.

request scope Any objects created in the *request scope* will be available as long as the request object is available. For example, if the JSP page uses an jsp:forward tag, then the bean should be applicable in the forwarded JSP also, if the scope defined is of *Request scope*.

session scope In JSP terms, the data associated with the user has *session scope*. A session does not correspond directly to the user, rather, it corresponds with a particular period of time the user spends at a site. Typically, this period is defined as, all the visits a user makes to a site between starting and exiting his browser.

application scope A bean with an *application scope* lives throughout the life of the JSP container itself. It is available from any page in the application.

class="package.class" Instantiates a bean from a class, using the new keyword and the class constructor. The class must not be abstract and must have a public, no-argument constructor. The package and class names are case-sensitive.

type="package.class" If the bean already exists in the scope, gives the bean a data type other than the class from which it was instantiated. The value of type must be a superclass of class or an interface implemented by class.

If you use type without class or beanName, no bean is instantiated. The package and class name are case-sensitive.

class="package.class" type="package.class" Instantiates a bean from the class named in package.class and assigns the bean the data type you specify in type. The value of type can be the same as class, a superclass of class, or an interface implemented by class. The class you specify in class must not be abstract and must have a public, no-argument constructor. The package and class names you use with both class and type are case-sensitive.

beanName="{package.class | <%=expression%>}"type="package.class" Instantiates a bean from a class, a serialized template, or an expression that evaluates to a class or serialized template. When you use beanName,the bean is instantiated by the java.beans.Beans.instantiate method. The Beans.instantiate method checks whether the package and class you specify represents a class or a serialized template. If they represent a serialized template, Beans.instantiate reads the serialized form (which has a name like package.class.ser) using a class loader.

The value of type can be the same as beanName, a superclass of beanName, or an interface implemented by beanName. The package and class names you use with both beanName and type are case-sensitive.

10.2.2 jsp:setProperty

The jsp:setProperty element sets the value of one or more properties in a bean, using the bean's setter methods. You must declare the bean with <jsp:useBean> before you set a property value with jsp:setProperty. Because jsp:useBean and jsp:setProperty work together, the bean instance names they use must match (that is, the value of name in jsp:setProperty and the value of id in jsp:useBean must be the same).

You can use jsp:setProperty to set property values in several ways.

- By passing all of the values the user enters (stored as parameters in the request object) to matching properties in the bean
- By passing a specific value the user enters to a specific property in the bean
- By setting a bean property to a value you specify as either a String or an expression that is evaluated at runtime

Each method of setting property values has its own syntax, as described in the next section.

JSP syntax The JSP syntax of jsp:setProperty is given below.

```
<jsp:setPropertyname="beanInstanceName"
{
 property="*"|
 property="propertyName"[param="parameterName"]|
 property="propertyName"value="{stringLiteral|
  '${' Expression '}'|<%= expression %>}"
}
/>
```

XML syntax The XML syntax of jsp:setProperty is given below.

```
<jsp:setPropertyname="beanInstanceName"
```

```
{
 property="*"|
 property="propertyName"[param="parameterName"]|
property="propertyName" value="{stringLiteral|     '${'Expression '}'| %=
    expression %}"
}
/>
```

Attributes The Attributes of jsp:setProperty are listed below.

name="beanInstanceName" The name of an instance of a bean that has already been created or located with a jsp:useBean element. The value of name must match the value of id in jsp:useBean. The jsp:useBean element must appear before jsp:setProperty in the JSP page.

property=""* Stores all of the values of request parameters in bean properties. The names of the bean properties must match the names of the request parameters. A bean property is usually defined by a variable declaration with matching getter and setter methods.

property="propertyName" [param="parameterName"] Sets one bean property to the value of one request parameter. In the syntax, property specifies the name of the bean property and param specifies the name of the request parameter by which data is being sent from the client to the server.

If the bean property and the request parameter have different names, you must specify both property and param. If they have the same name, you can specify property and omit param.

If a parameter has an empty or null value, the corresponding bean property is not set.

property="propertyName" value="{string | <%= expression %>}" Sets one bean property to a specific value. The value can be a String or an expression that is evaluated at runtime.

Type conversions for <jsp:setProperty>

Property type	Conversion to String
boolean or Boolean	java.lang.Boolean.valueOf(String)
byte or Byte	java.lang.Byte.valueOf(String)
char or Character	java.lang.Character.valueOf(String)
double or Double	java.lang.Double.valueOf(String)
integer or Integer	java.lang.Integer.valueOf(String)
float or Float	java.lang.Float.valueOf(String)
long or Long	java.lang.Long.valueOf(String)

10.2.3 jsp:getProperty

The jsp:getProperty element gets a bean property value using the property's getter methods and inserts the value into the response. You must create or locate a bean with <jsp:useBean> before you use jsp:getProperty.

The jsp:getProperty element has a few limitations you should be aware of:

- You cannot use jsp:getProperty to retrieve the values of an indexed property.

You can use jsp:getProperty with JavaBeans components, but not with enterprise beans. As alternatives, you can write a JSP page that retrieves values from a bean that in turn retrieves values from an enterprise bean, or you can write a custom tag that retrieves values from an enterprise bean directly.

JSP syntax The JSP syntax of jsp:getProperty is given below.

<jsp:getPropertyname="beanInstanceName"property="propertyName"/>

Attributes The attributes of jsp:getProperty is listed below.

name="beanInstanceName" The name of an object (usually an instance of a bean) as declared in a <jsp:useBean> element.

property="propertyName" The name of the bean property whose value you want to display. The property is declared as a variable in a bean and must have a corresponding getter method.

Type conversions for <jsp:getProperty>

Property type	Conversion to String
Boolean	java.lang.Boolean.toString(boolean)
byte	java.lang.Byte.toString(byte)
char	java.lang.Character.toString(char)
double	java.lang.Double.valueOf(String)
int	java.lang.Integer.toString(int)
float	java.lang.Float.toString(float)
long	java.lang.Long.toString(long)

Note If you use jsp:getProperty to retrieve a property value that is null, a NullPointerException is thrown. However, if you use a scriptlet or expression to retrieve the value, the string null is displayed in the browser; see Scriptlet or Expression for more information.

Example 10.2

Write a bean class whose code is given below. The code has one public class called doubleInt that returns an integer as a result of multiplication of the argument by 2. Compile this bean class and copy under WEB-INF.

CalculatorBean.jsp

```
packageankur;
publicclassCalculatorBean{
publicintdoubleIt(int number){
return2*number;
}
}
```

bean.jsp

```
<jsp:usebeanid="thebean"class="ankur.calculatorbean"/>
<html>
```

```
<head>
</head>
<body>
<%
inti=4;
 intj=thebean.doubleit(i);
 out.print("2*4="+j);
%>
</body>
</html>
```

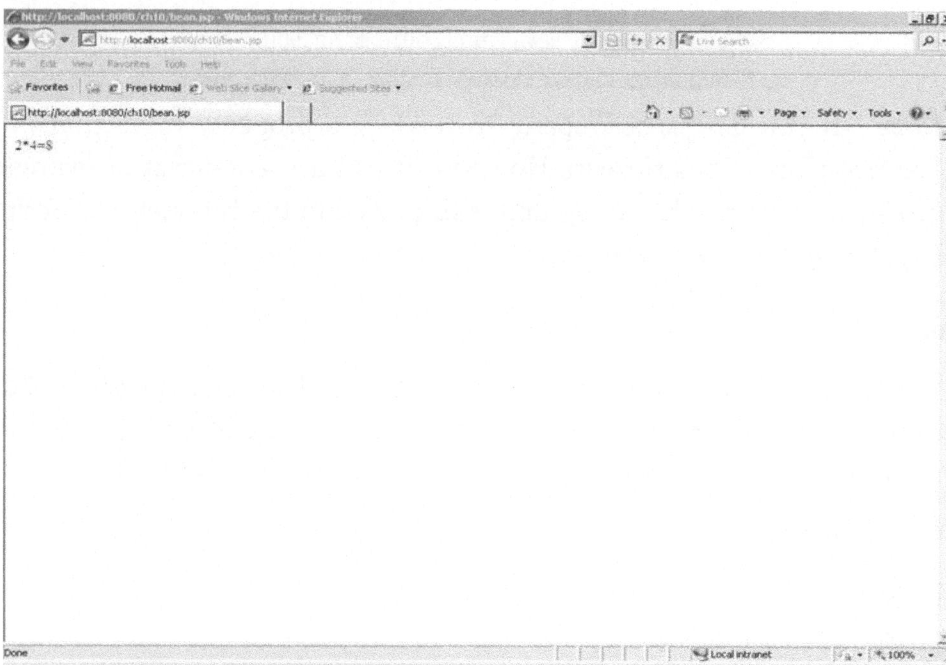

Figure 10.1 Output of bean.jsp

This screen (Figure 10.1) display the multiplication result of two integer using Bean.

Example 10.3

The following example demonstrates the use of the jsp:getProperty and jsp:setProperty.

CalcBean.java

```
packageankur;
publicclassCalcBean{
 privateintmemory;
  publicvoidsetMemory(int number){
   memory=number;
  }
   publicintgetMemory(){
   return memory;
  }
  publicintdoubleIt(int number){
  return2*number;
 }
}
```

bean1.jsp: accessing the property using jsp:setProperty and jsp:getProperty

```
<jsp:useBeanid="theBean"class="ankur.CalcBean"/>
<jsp:setPropertyname="theBean"property="memory"value="200"/>
Thevalueofmemoryis<jsp:getPropertyname="theBean"property="memory"/>
```

Example 10.4

bean2.jsp: Calling accessor method like ordinary methods

```
<jsp:useBeanid="theBean"class="ankur.CalcBean"/>
< %
 theBean.setMemory(900);
% >
Thevalueofmemoryis<%=theBean.getMemory()%>
```

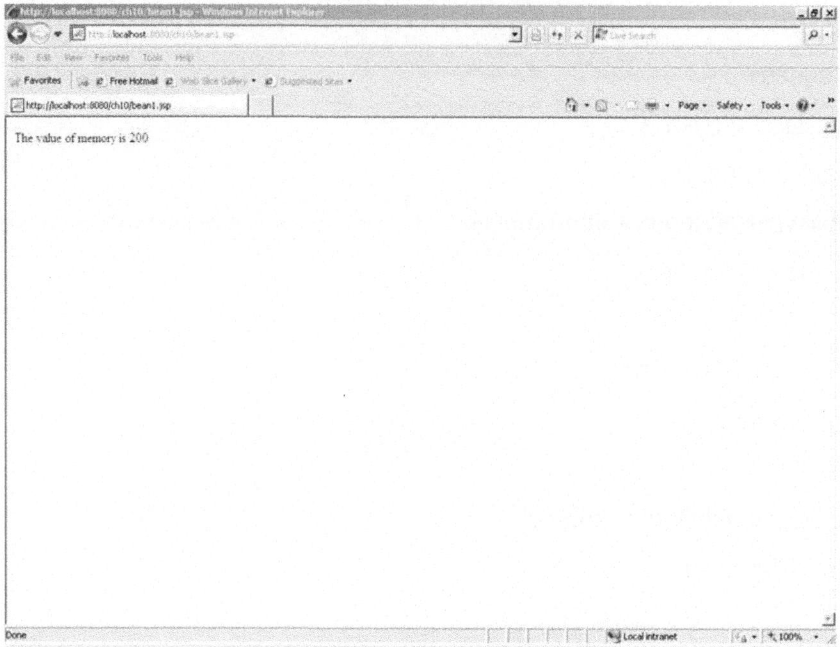

Figure 10.2 Output of bean 1.jsp

This screen (Figure 10.2) display the value of memory.

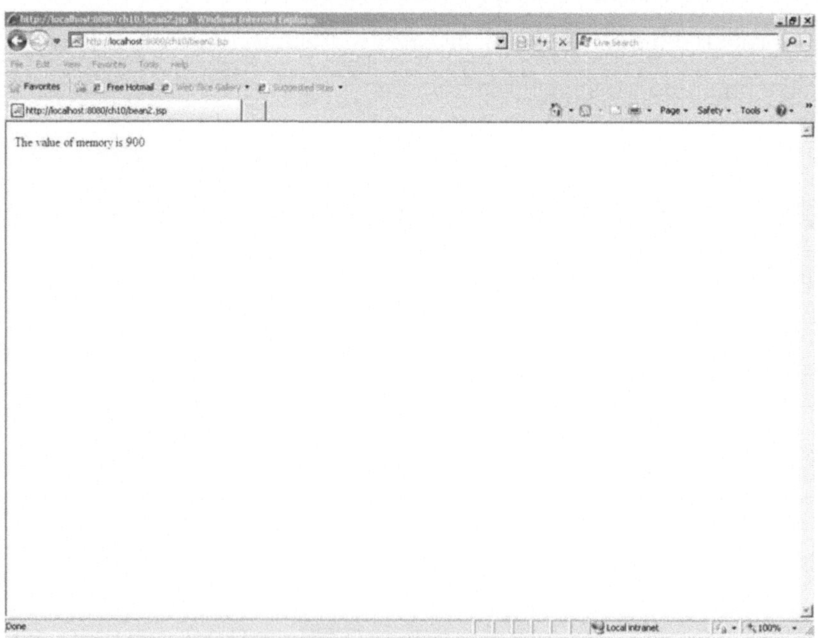

Figure 10.3 Output of bean 2.jsp

This screen (Figure 10.3) display the value of memory by accesor method.

Example 10.5

bean3.jsp: Feeding a JSP Expression on the value Attribute

```
<jsp:useBeanid="theBean"class="ankur.CalcBean"/>
<%
 inti=2;
%>
<jsp:setPropertyname="theBean"property="memory" value="<%= 20*i %>"/>
Thevalueofmemoryis<jsp:getPropertyname="theBean"property="memory"/>
```

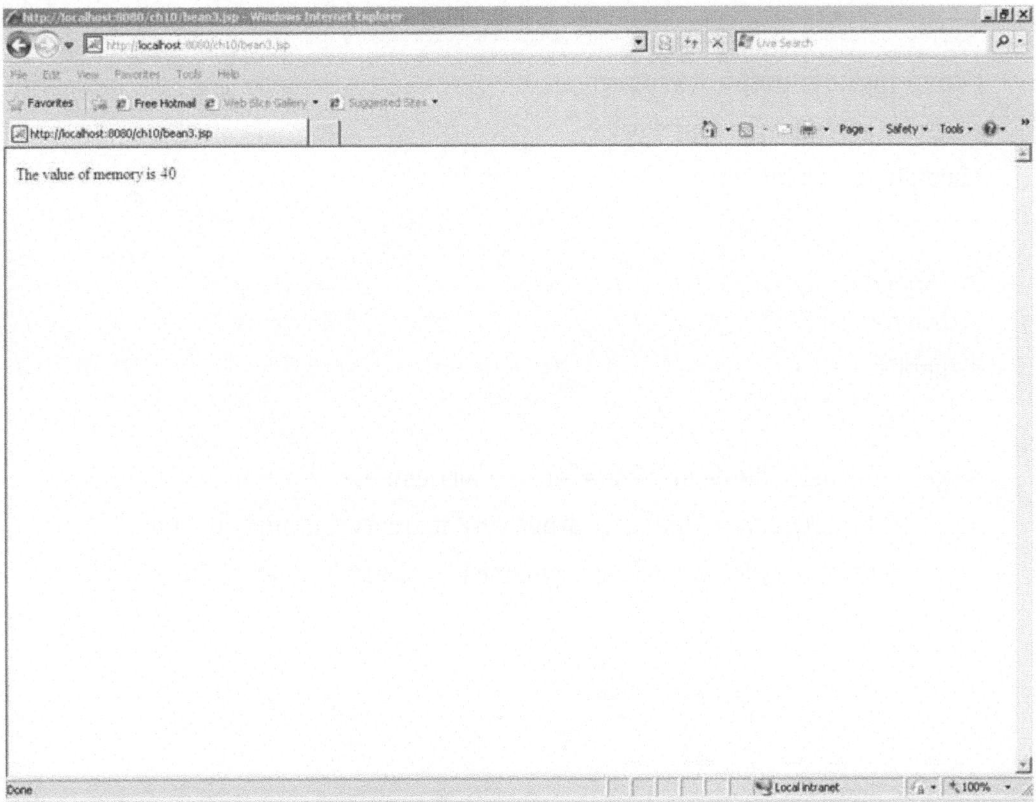

Figure 10.4 Output of bean 3.jsp

This screen (Figure 10.4) display the value of memory using expression.

Example 10.6 setting value from a Request

This example show setting value from a Request

form.jsp:

```
<html>
<head>
<title>passing a value</title>
</head>
<body>
<center>
pleasetypeinanumberinthebox
<br>
<form method=post action=page.jsp>
<inputtype=textname=memory>
<inputtype=submit>
</form>
</center>
</body>
</html>
```

page.jsp

```
<jsp:useBeanid="theBean"class="ankur.CalcBean"/>
<jsp:setPropertyname="theBean"property="memory"param="memory"/>
Thevalueofmemoryis<jsp:getPropertyname="theBean"property="memory"/>
```

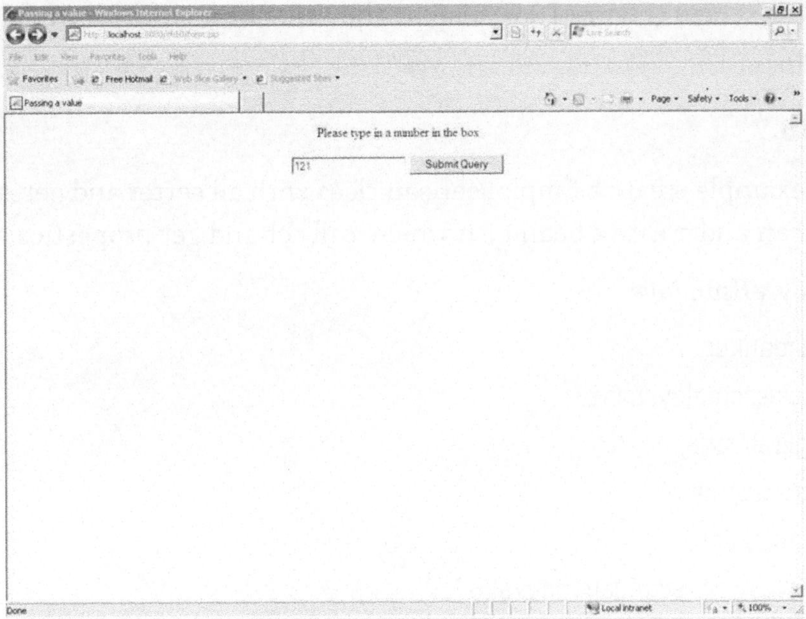

Figure 10.5 Output of form.jsp

This screen (Figure 10.5) display text box now enter a value and press submit query button.

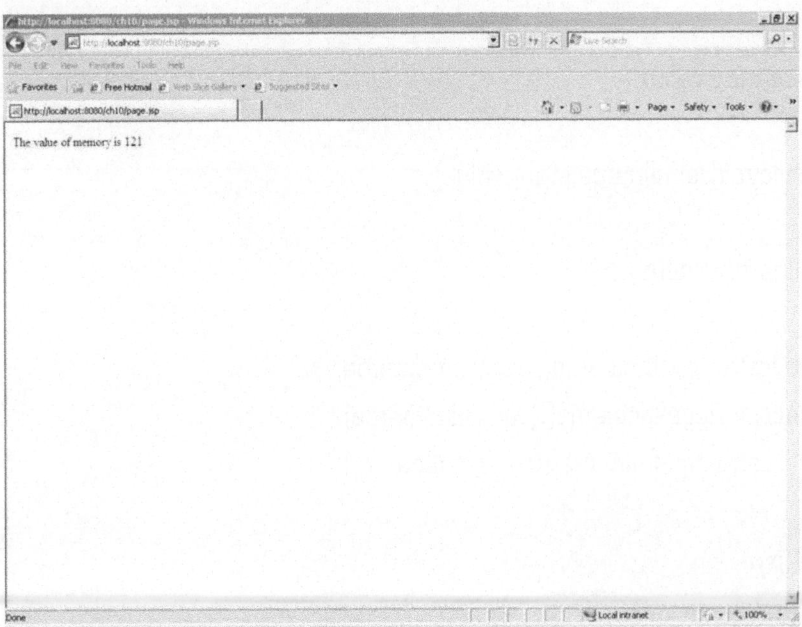

Figure 10.6 Output of page.jsp

This screen (Figure 10.6) display the value of memory entered by previous screen.

Example 10.7

This example create a EmployeeBean class with all setter and getter method of the bean and create a bean4 jsp which call set and get properties of the bean.

EmployeeBean.java

```
packageankur;
publicclassemployeebean{
 stringfirstname;
 stringlastname;
 stringfullname;
 publicvoidsetfirstname(stringvalue)
  {
  firstname=value;
    }
    publicvoidsetlastname(stringvalue)
    {
    lastname=value;
    }
    publicvoidsetfullname(stringvalue)
    {
     fullname=value;
    }
    publicstringgetfirstname(){returnfirstname;}
    publicstringgetlastname(){returnlastname;}
    publicstringgetfullname(){returnfullname;}
}
```

bean4.jsp

```
<head>
```

```
<title>example:simplejavabean</title>
<jsp:usebeanid="employeebean"scope="page"
class="ankur.employeebean"/>
</head>
<body>
<%
 stringlast="saxena";
%>
<%--setbeanproperties--%>
<jsp:setpropertyname="employeebean"property="firstname"
     value="ankur"/>
<jsp:setpropertyname="employeebean"property="lastname"
     value="<%=last%>"/>
<jsp:setpropertyname="employeebean"property="fullname"  value="ankursaxena"/
     >
<%--getbeanproperties--%>
<p>
</p>
firstname:<%=employeebean.getfirstname()%> <br>
lastname:<%=employeebean.getlastname()%> <br>
fullname:<jsp:getpropertyname="employeebean"property="fullname"/>
</body>
</html>
```

Note, we can also call the associated property methods directly within Java code using syntax similar to,

```
employeeBean.setFirstName("Ankur");
employeeBean.setLastName("Saxena");
Stringname=employeeBean.getFullName();
```

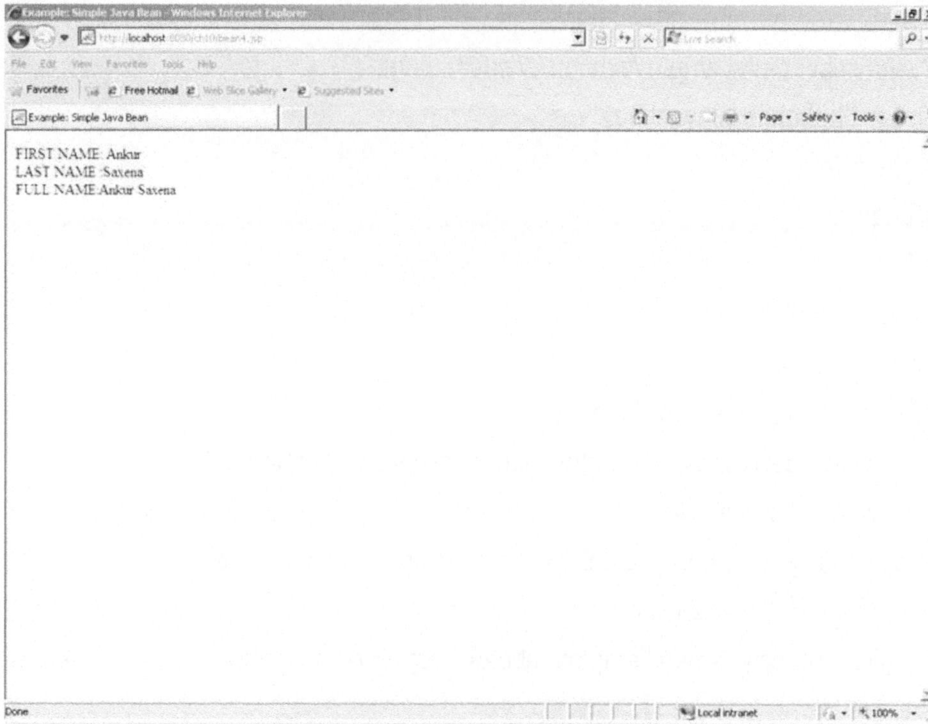

Figure 10.7 Output of bean 4.jsp

This screen (Figure 10.7) display the value of FIRST NAME,LAST NAME,FULL NAME using set and get bean properties.

Example **10.8** *Beans and Form processing*

Forms are a very common method of interactions in websites. JSP makes forms processing specially easy.

The standard way of handling forms in JSP is to define a "bean". This is not a full Java bean. You just need to define a class that has a field corresponding to each field in the form. The class fields must have "setters" that match the names of the form fields. For instance, let us modify our **get.html** to also collect email address and age.

get.html

```
<html>
<body>
```

```
<formmethod=postaction="new.jsp">
what'syourname?<input type=textname=usernamesize=20> <br>
what'syoure-mailaddress?<inputtype=text name=email size=20> <br>
what'syourage?<inputtype=textname=agesize=4>
<p> <input type=submit>
</form>
</body>
</html>
```

To collect this data, we define a Java class with fields "username", "email" and "age" and we provide setter methods "setUsername", "setEmail" and "setAge", as shown. A "setter" method is just a method that starts with "set" followed by the name of the field. The first character of the field name is upper-cased. So if the field is "email", its "setter" method will be "setEmail". “Getter” methods are defined similarly, with "get" instead of "set". Note that the setters and getters must be public.

```
packageankur;
publicclassUserData{
Stringusername;
Stringemail;
intage;
publicvoidsetUsername(Stringvalue )
 {
 username=value;
  }
  publicvoidsetEmail(Stringvalue)
  {
email=value;
 }
  publicvoidsetAge(intvalue)
 {
age=value;
```

```
  }
publicStringgetUsername(){returnusername;}
publicStringgetEmail(){returnemail;}
publicintgetAge(){returnage;}
}
```

Now let us change **"new.jsp"** to use a bean to collect the data.

```
<jsp:usebeanid="user"class="ankur.userdata"scope="session"/>
    <jsp:setpropertyname="user"property="*"/>
<html>
<body>
<ahref="nextpage.jsp">continue</a>
</body>
</html>
```

All we need to do now is to add the jsp:useBean tag and the jsp:setProperty tag. The useBean tag will look for an instance of the ankur.UserData in the session. If the instance is already there, it will update the old instance. Otherwise, it will create a new instance of ankur.UserData (the instance of the ankur.UserData is called a bean), and put it in the session.

The setProperty tag will automatically collect the input data, match names against the bean method names, and place the data in the bean.

Let us modify NextPage.jsp to retrieve the data from bean..

```
<jsp:usebeanid="user"class="ankur.userdata"scope="session"/>
<html>
<body>
youentered<br>
name:<%=user.getusername()%> <br>
email:<%=user.getemail()%> <br>
age:<%=user.getage()%> <br>
</body>
</html>
```

Notice that the same useBean tag is repeated. The bean is available as the variable named user of class ankur.UserData. The data entered by the user is all collected in the bean.

We do not actually need the new.jsp, the target of get.html could have been NextPage.jsp, and the data would still be available the same way as long as we added a jsp:setProperty tag. But in the next tutorial, we will actually use new.jsp as an error handler that automatically forwards the request to NextPage.jsp, or asks the user to correct the erroneous data.

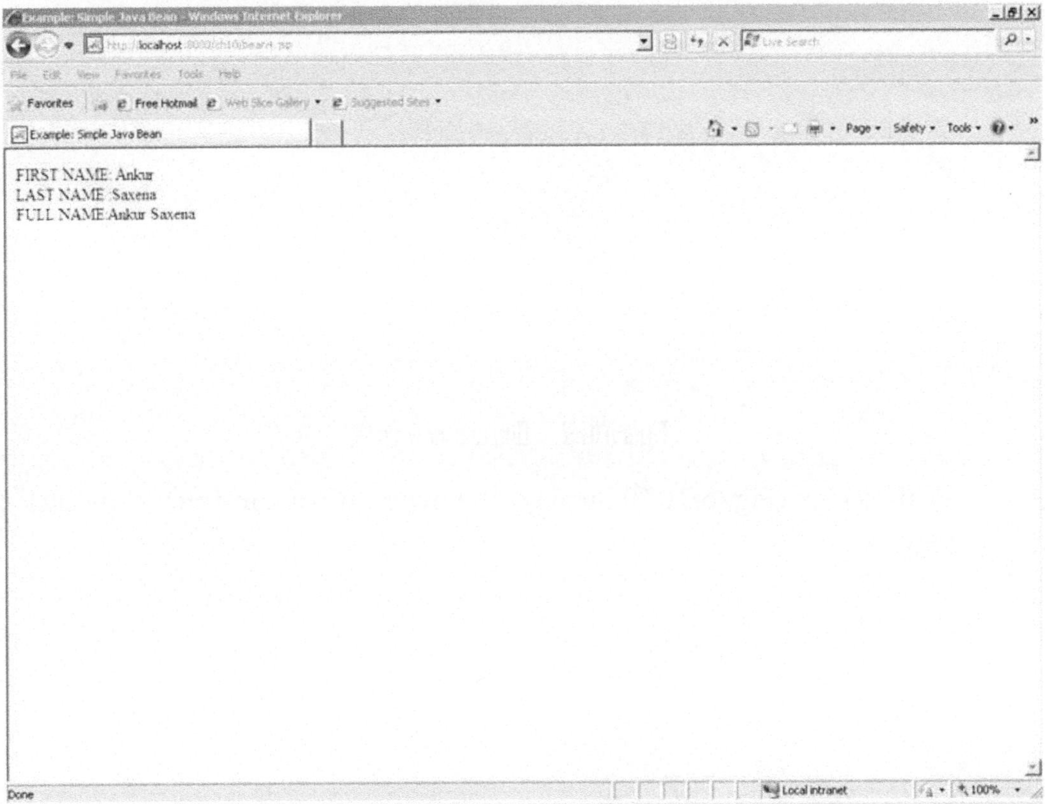

Figure 10.8 Output of get.html

This screen (Figure 10.8) display three text box. Now enter name, email address, age and press submit query button.

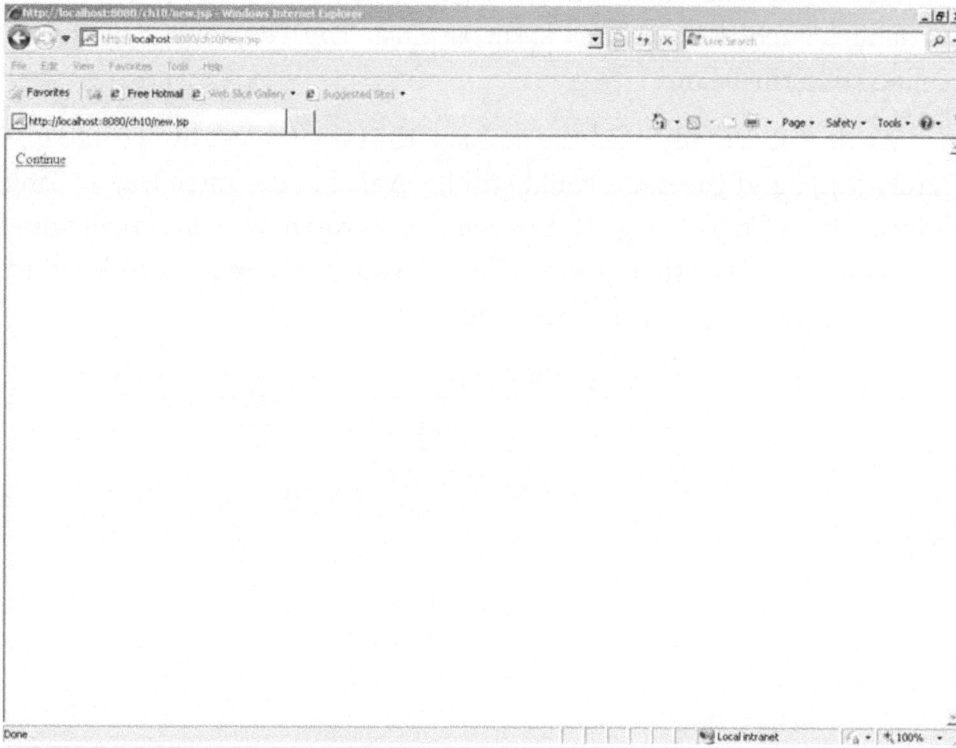

Figure 10.9 Output of new.jsp

This screen (Figure10.9) displays the hyperlink of continue. Now click this hyperlink.

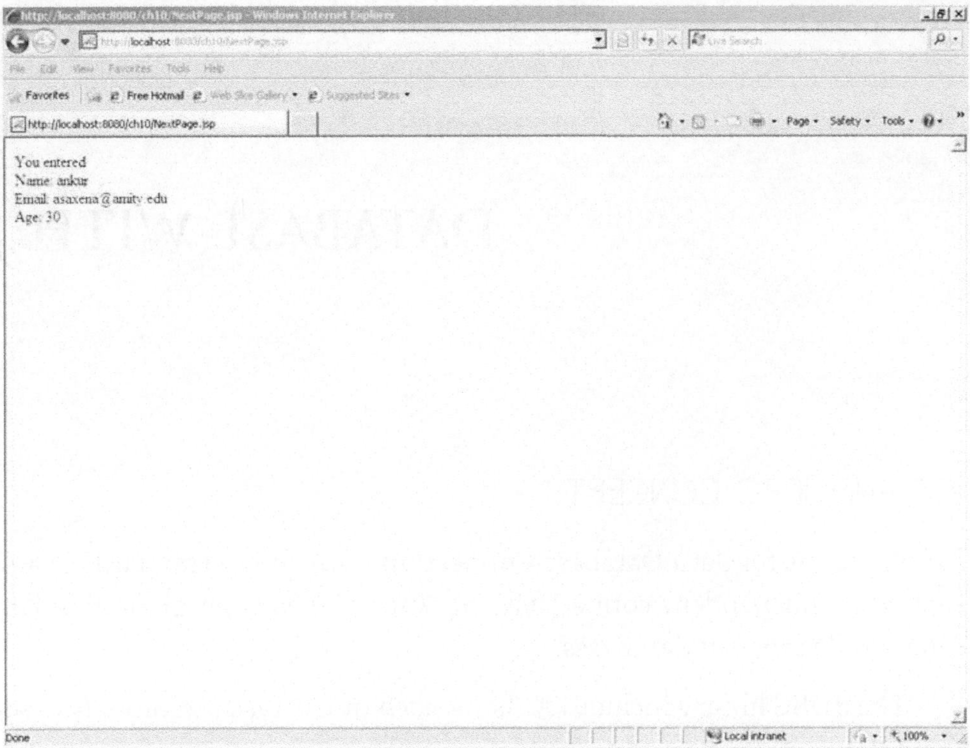

Figure 10.10 Output of Nextpage.jsp

This screen (Figure 10.10) display the value of name, email and age entered by previous screen.

REVIEW QUESTIONS

1. What is bean container?
2. What is the advantage of java bean?
3. How to compile a java bean?
4. Explain type conversions for <jsp:setProperty> and <jsp:getProperty>
5. What is bean? Explain <jsp:UseBean> action.
6. What is the difference between <jsp:setProperty> and <jsp:getProperty> actions.

11
DATABASE WITH JSP

11.1 JSP AND JDBC CONCEPTS

JDBC stands for Java Database Connectivity, which is a standard Java API for database-independent connectivity between the Java programming language and a wide range of databases.

The JDBC library includes APIs for each of the tasks commonly associated with database usage.

- Making a connection to a database
- Creating SQL or MySQL statements
- Executing that SQL or MySQL queries in the database
- Viewing & Modifying the resulting records

The following code fragment shows these steps in actions.

//Making a connection to a database:

```
connectioncon=DriverManager.getConnection("jdbc:odbc:HRM","admin","password");
```

//Creating SQL or MySQL statements

```
statementstat = con.createStatement();
```

//Executing that SQL or MySQL queries in the database

```
Stringemp="select*fromemployee";
ResultSet resultset = statement.executeQuery(emp);
```

//Viewing & Modifying the resulting records

```
while(resultset.next())
{
out.print(+rs.getString(1));
}
```

11.2 JDBC ARCHITECTURE

The JDBC API supports both two-tier and three-tier processing models for database access, but in general JDBC architecture consists of two layers.

1. **JDBC API** This provides the application-to-JDBC Manager connection.
2. **JDBC Driver API** This supports the JDBC Manager-to-Driver connection.

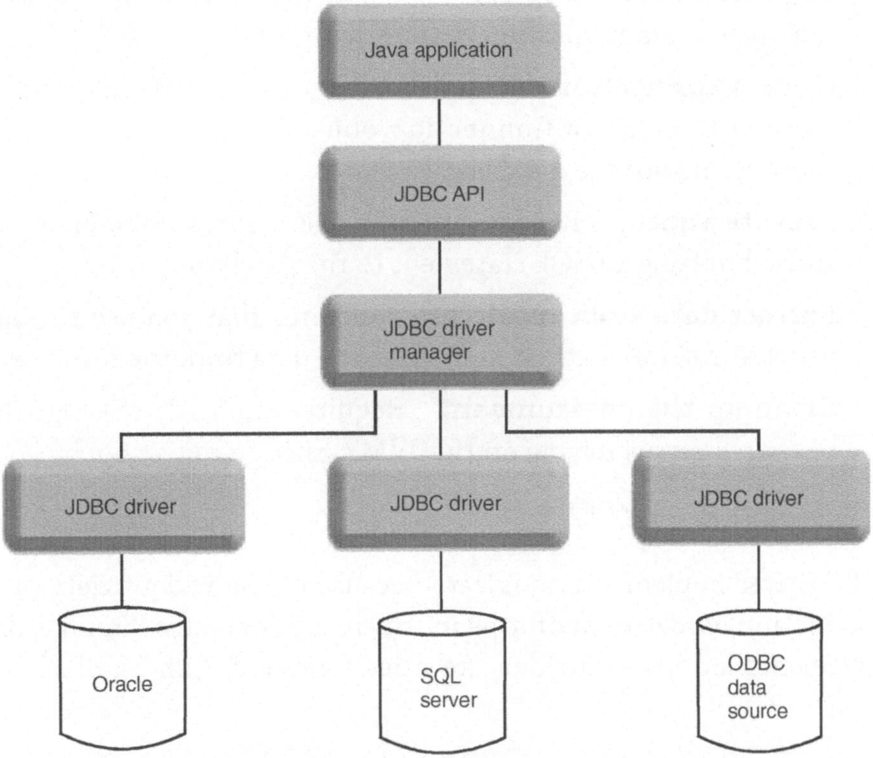

Figure 11.1 JDBC a rchitecture diagram

The JDBC API uses a driver manager and database-specific drivers to provide transparent connectivity to heterogeneous databases.

The JDBC driver manager ensures that the correct driver is used to access each data source. The driver manager is capable of supporting multiple concurrent drivers connected to multiple heterogeneous databases.

Figure 11.1 shows the architectural diagram, which shows the location of the driver manager with respect to the JDBC drivers and the Java application.

11.2.1 Creating JDBC Application

There are six steps involved in building a JDBC application. They are:

1. **Import the packages** Requires that you include the packages containing the JDBC classes needed for database programming. Most often, using import java.sql.* will suffice.
2. **Register the JDBC driver** Requires that you initialize a driver so you can open a communication channel with the database.
3. **Open a connection** Requires using the DriverManager.getConnection() method to create a Connection object, which represents a physical connection with the database.
4. **Execute a query** Requires using an object of type Statement for building and submitting an SQL statement to the database.
5. **Extract data from result set** Requires that you use the appropriate ResultSet.getXXX() method to retrieve the data from the result set.
6. **Clean up the environment** Requires explicitly closing all database resources versus relying on the JVM's garbage collection.

11.3 JDBC DRIVERS TYPES

JDBC driver implementations vary because of the wide variety of operating systems and hardware platforms in which Java operates. Sun has divided the implementation types into four categories, types 1, 2, 3, and 4, which is explained below.

11.3.1 Type 1 JDBC-ODBC Bridge Driver

The first of JDBC driver is JDBC–ODBC Bridge. This driver is used to access ODBC drivers installed on each client machine. Using ODBC requires configuring on your system a Data Source Name (DSN) that represents the target database.

When Java first came out, this was a useful driver because most databases only supported ODBC access but now this type of driver is recommended only for experimental use or when no other alternative is available. The Figure 11.2 shows type 1 architecture.

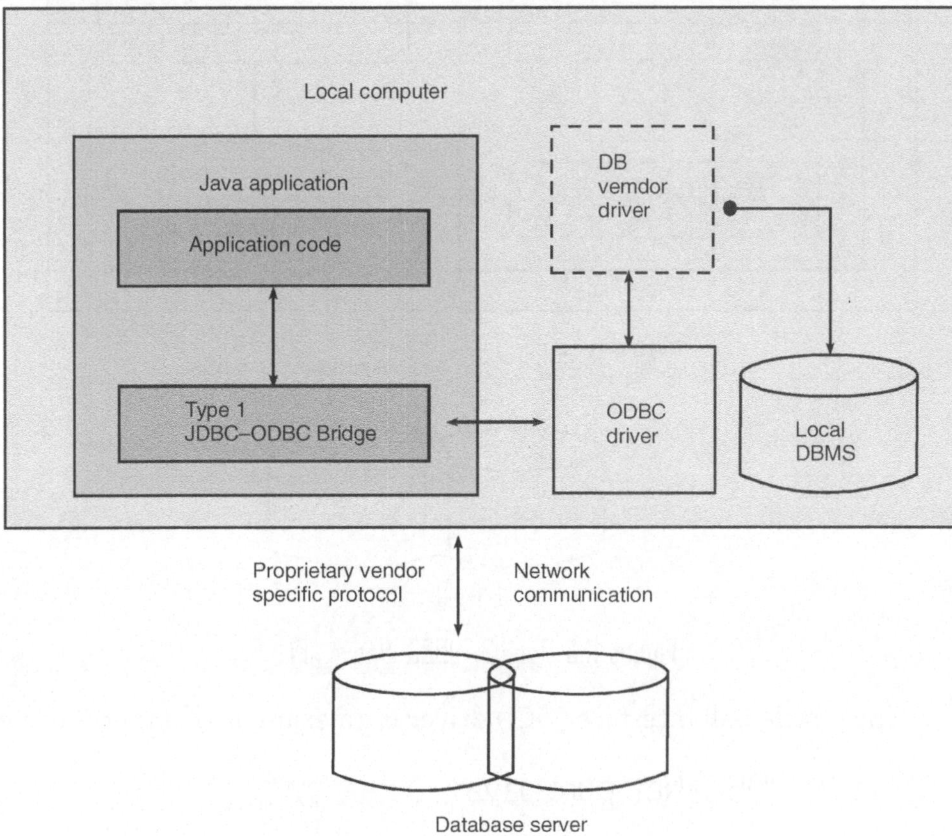

Figure 11.2 Type 1—JDBC-ODBC bridge driver

11.3.2 Type 2 JDBC-Native API

JDBC API calls convert JDBC commands into DBMS-specific native calls which are unique to the database. These drivers typically provided by the database vendors are used in the same manner as the JDBC–ODBC bridge, the vendor-specific driver must be installed on each client machine. The Figure 11.3 shows the type 2 Architecture.

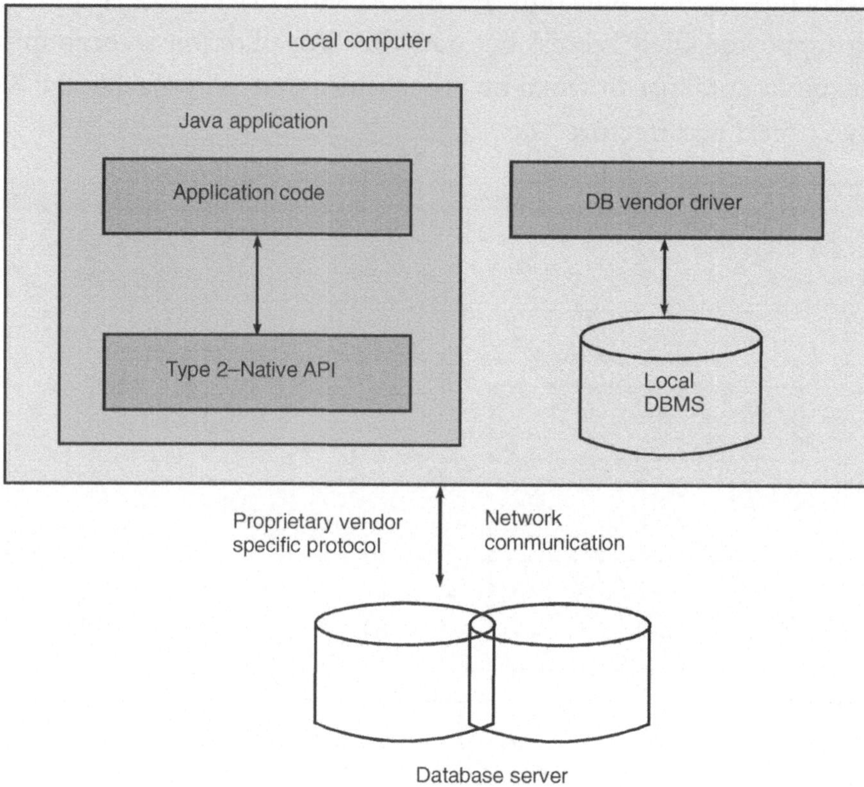

Figure 11.3 Type 2—JDBC-Native API

The Oracle Call Interface (OCI) driver is an example of a type 2 driver.

11.3.3 Type 3 JDBC-Net pure Java

The JDBC—Net a three-tier approach is used for accessing databases. The JDBC clients use standard network sockets to communicate with a middleware application server. The socket information is then translated by the middleware application server into the call format required by the DBMS, and forwarded to the database server.

This kind of driver is extremely flexible, since it requires no code installed on the client and a single driver can actually provide access to multiple databases. Figure 11.4 shows the type 3 architecture.

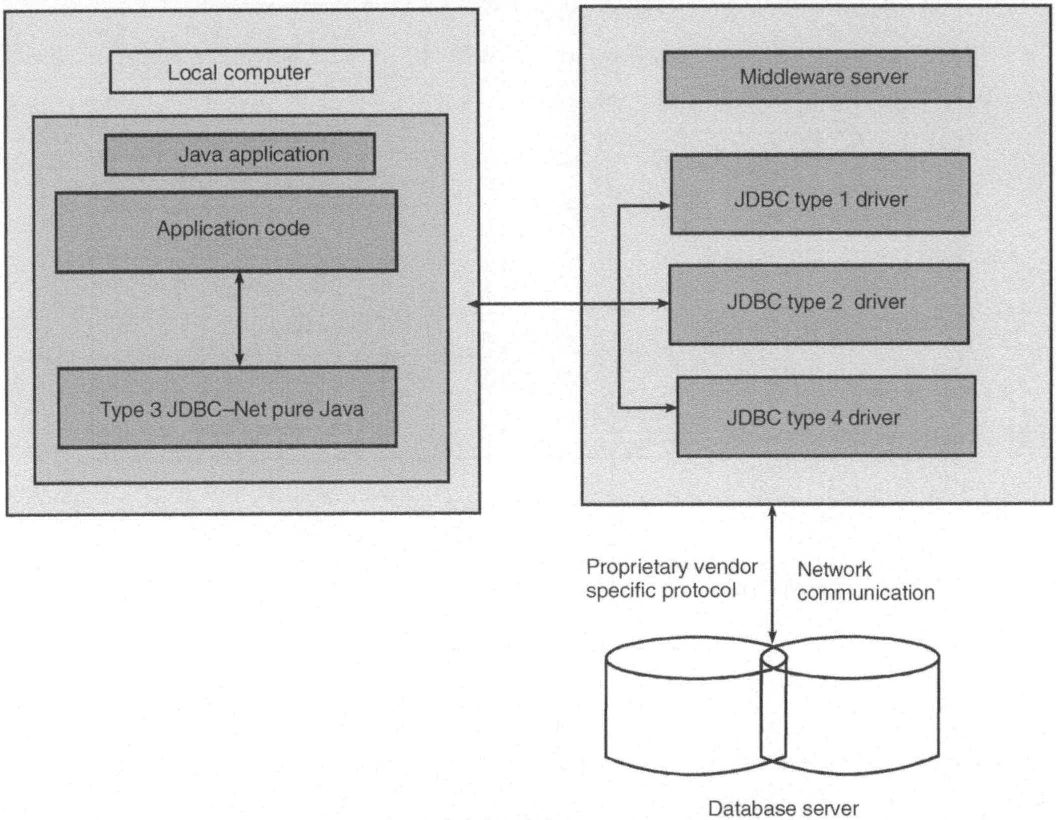

Figure 11.4 Type 3– JDBC -Net Pure Java

You can think of the application server as a JDBC "proxy," meaning that it makes calls for the client application. As a result, you need some knowledge of the application server's configuration in order to effectively use this driver type.

11.3.4 Type 4 Native -Protocol, pure Java Driver

Type 4 driver is a pure Java-based driver that communicates directly with vendor's database through socket connection. This is the highest performance driver available for the database and is usually provided by the vendor itself.

This kind of driver is extremely flexible, you do not need to install special software on the client or server. Further, these drivers can be downloaded dynamically. Figure 11.5 shows the type 4 Architecture.

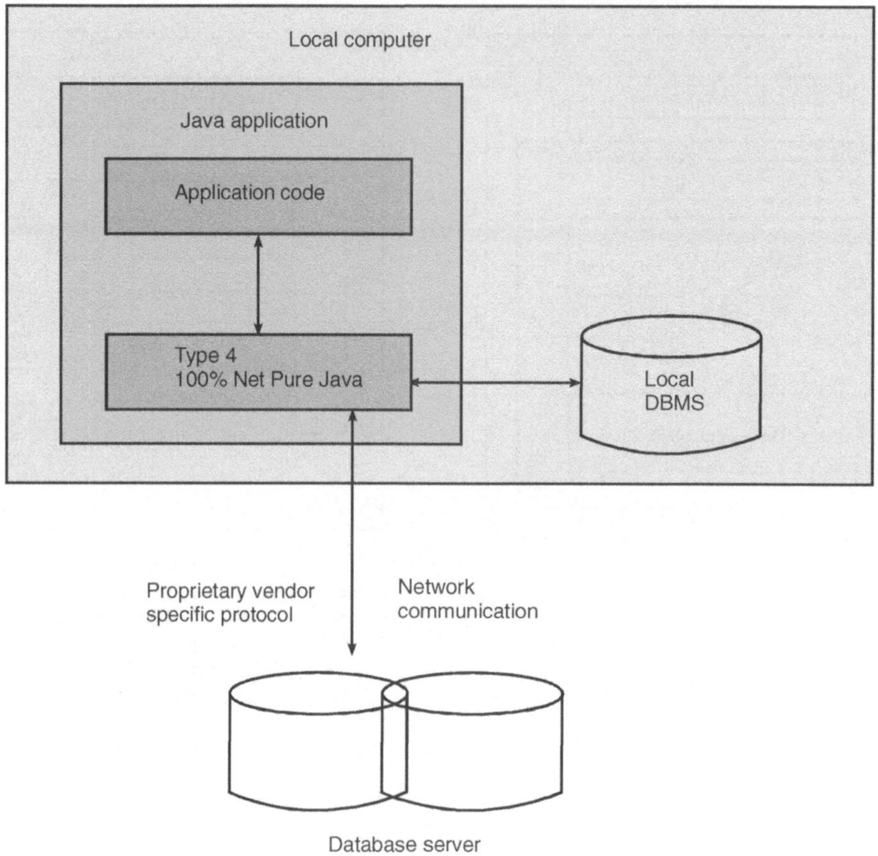

Figure 11.5 Type 4 Pure Java based driver

11.3.5 Which Driver Should be Used

Driver types are used to categorize the technology used to connect to the database. A JDBC driver vendor uses these types to describe how their product operates. Some JDBC driver types are better suited for some applications than others. The selection of drivers depend upon application requirement.

- If you are accessing one type of database, such as Oracle, Sybase, or IBM, the preferred driver type is 4.
- If your Java application is accessing multiple types of databases at the same time, type 3 is the preferred driver.
- Type 2 drivers are useful in situations where a type 3 or type 4 driver is not available yet for your database.

- The type 1 driver is not considered a deployment-level driver and is typically used for development and testing purposes only.
- Once a connection is obtained we can interact with the database. The JDBC Statement, CallableStatement, and PreparedStatement interfaces define the methods and properties that enable you to send SQL or PL/SQL commands and receive data from your database.
- They also define methods that help bridge data type differences between Java and SQL data types used in a database.

Table 11.1 provides a summary of each interface's purpose to understand how you decide which interface to use.

Table 11.1 Purpose of JDBC interface

JDBC support class or interfaces	Description
java.sql.DriverManager	This interface manages a list of database drivers. Matches connection requests from the java application with the proper database driver using communication subprotocol. The first driver that recognizes a certain subprotocol under JDBC will be used to establish a database connection.
java.sql.Connection	Interface with all methods for contacting a database. The connection object represents communication context, i.e., all communication with database is through connection object only.
java.sql.Statement	Used for general-purpose access to your database. Useful when you are using static SQL statements at runtime. The Statement interface cannot accept parameters.
java.sql.preparedStatement	Use when you plan to use the SQL statements many times. The preparedStatement interface accepts input parameters at runtime.

Table 11.1 (continues)

JDBC support class or interfaces	Description
java.sql .CallableStatement	Use when you want to access database stored procedures. The CallableStatement interface can also accept runtime input parameters.
java.sql.ResultSet	These objects hold data retrieved from a database after you execute an SQL query using statement objects. It acts as an iterator to allow you to move through its data.
java.sql.SQLException	This class handles any errors that occur in a database application.
java.sql.SQLWarning	Extends SQLException, provides extra information on database access errors.

11.4 HOW JSP AND JDBC FIT TOGETHER

The classic JDBC deployment is illustrated in the Figure below. JSP can readily accommodate this scenario, although the scheme takes no advantage of the flexibility offered by JSP. Figure 11.6 shows the JDBC deployment.

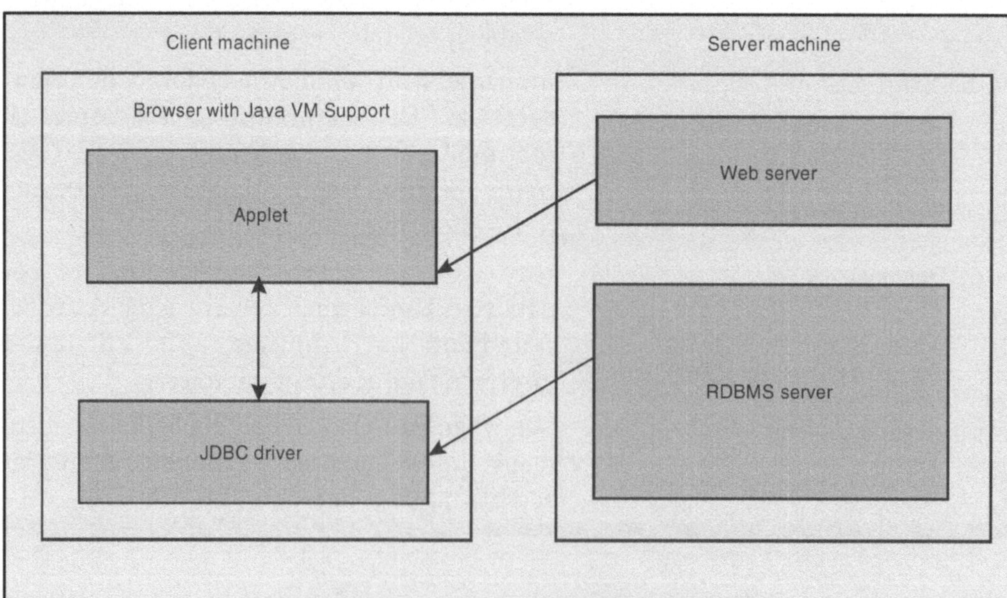

Figure 11.6 Classic JDBC deployment

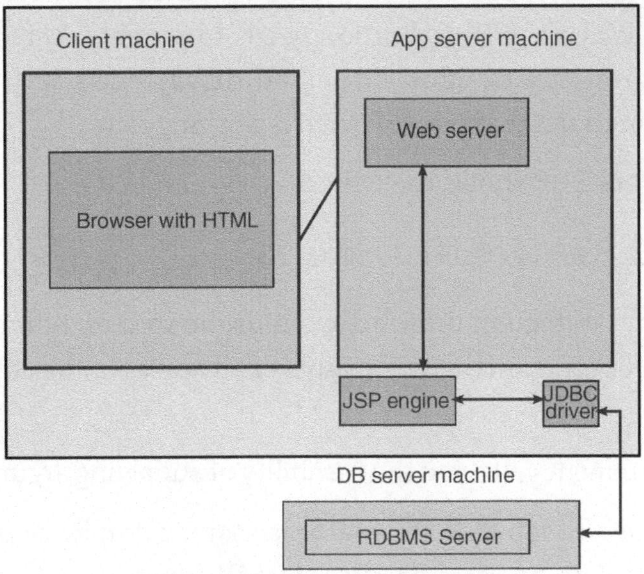

Figure 11.7 JDBC API Architecture

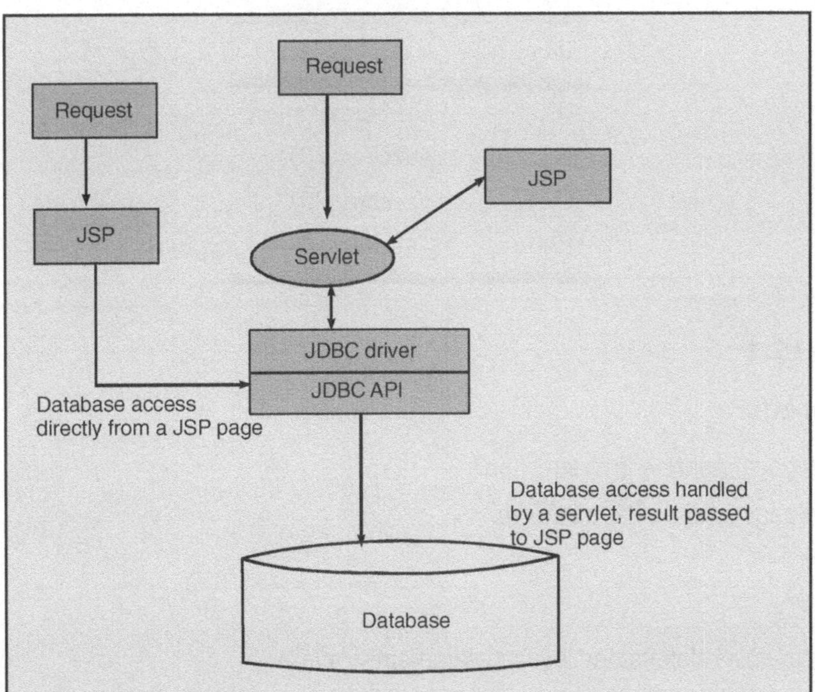

Figure 11.8 Database access option in JSP

The designers of JSP choose to leverage Java's powerful, popular, database API-JDBC. When a JSP application needs to communicate with a database, it does so through a vendor provided driver class written to the JDBC API.Accessing a database in JSP then is nothing new.

The Figure 11.8 shows the DB Access option in JSP.

11.5 PREPARED STATEMENT USING JSP

The Prepared statement interface extends the statement interface which gives you added functionality with a couple of advantages over a generic statement object.

This statement gives you the flexibility of supplying arguments dynamically.

Create a database First create a database named 'employee' in MySQL and table named "emp_info" in same database by SQL query given in Table 11.2.

Table 11.2 emp_info

Field	Type
ID	int primary key
Name	varchar(20)
City	varchar(20)
Phone	varchar(10)

***Example* 11.1**

prepare.jsp

```
<%@pageimport="java.sql.*" %>
<%@pageimport="java.io.*" %>
<html>
<head>
  <title>insertdatausingpreparedstatement</title>
</head>
 <body>
```

```
<br>welcomeinjavaserverpages
<formaction="prepare.jsp"method="get">
<table>
 <tr>
<th>name</th>
<td> <input type="text" name="name"> </td>
 </tr>
 <tr>
  <th>city</th>
  <td> <input type="text" name="city"> </td>
   </tr>
  <tr>
   <th>phone</th>
      <td> <inputtype="text" name="phone"> </td>
</tr>
 <tr>
 <th> </th>
 <td> <input type="submit"value="submit"> </td>
 </tr>
 </table>
< %
  stringname=request.getparameter("name");
  stringcity=request.getparameter("city");
  stringphone=request.getparameter("phone");
  /*createstringofconnectionurlwithinspecified
  databasenameisemployee.*/
   stringconnectionurl="jdbc:mysql://localhost:3306/employee";
  //declareaconnectionbyusingconnectioninterface
   connection connection=null;
  //declareobjectofstatementinterfacethatusesfor
   executingsqlstatements.
```

```
preparedstatementpstatement= null;
//loadjbbcdriver"com.mysql.jdbc.driver"
class.forname("com.mysql.jdbc.driver").newinstance();
int updatequery =0;
//checkifthetextboxisempty
if(name!=null && city!=null && phone!=null){
// check if the text box having only blank spaces
if(name!="" && city!="" && phone!="") {
try {
/*createaconnectionbyusinggetconnection()methodthattakesparametersofstringtype
connectionurl,usernameandpasswordtoconnecttodatabase. */
connection = drivermanager.getconnection
(connectionurl, "root","root");
//sqlquerytoinsertvaluesinthesecifiedtable.
stringquerystring="insertintoemp_info(name,
address,phone)values(?, ?, ?)";
/*createstatement()isusedforcreatestatementobjectthatisusedforsendingsqlstatements to
thespecifieddatabase. */
pstatement=connection.preparestatement(querystring);
pstatement.setstring(1, name);
pstatement.setstring(2, city);
pstatement.setstring(3, phone);
updatequery = pstatement.executeupdate();
if (updatequery != 0) { %>
<br>
<table>
<tr> <th>data is inserted successfully
</th> </tr>
</table>
<%
}
```

```
    }
  catch (exception ex) {
  out.println("unable to connect to batabase.");
   }
  finally {
    // close all the connections.
   pstatement.close();
  connection.close();
   }
  }
 }
%>
</form>
</body>
</html>
```

11.6 CALLABLE STATEMENT USING JSP

Just as a connection object creates the statement and prepared statement objects, it also creates the callable statement object which would be used to execute a call to a database stored procedure.

Step 1 Write the procedure

```
createor replaceprocedure getempage
(idinnumber,ageoutnumber)
is
return_age number;
begin
selectageintoreturn_agefromempwhere id:=id;
return_age;
end;
```

Step 2 Java code to execute procedure

```
publicclasscalldb{
publicstaticvoidmain(string[] args){
try{
/** loading the driver*/
class.forname("com.oracle.jdbc.driver");
/** getting connection*/
connectioncon=drivermanager.getconnection("jdbc:oracle://localhost:3306/ank");
/** creating callablestatement*/
callablestatementcall = con.preparecall("call getempage(?,?)");
call.setint(1,24); //settingempid=24
call.registeroutparameter(2,types.integer);
call.execute();
system.out.println("emp age is "+call.getint(2));
stmt.close();
con.close();
}catch(exception e){
e.printstacktrace();
}
 }
}
```

11.7 ACCESS DATABASE USING BEAN

Create a database First create a database named 'ankur' and table named "ank" in same database and is shown in Table 11.3.

Table 11.3 ank

Field	Type
userName	varchar(20)
password	varchar(20)

The login page for this application is shown below.

Example 11.2

login.jsp

```
<%@pagelanguage="java" import="java.sql.*" %>
<html>
<body>
<formname="form1"method="post">
  username
   <inputtype="text"name="username"> <br>
  password<inputtype="text"name="message"> <br>
 <inputtype="submit"value="submit">
<jsp:usebeanid="abc"class="ankur.sqlbean"scope="page">
<jsp:setpropertyname="abc" property="*"/>
</jsp:usebean>
</form>
<% abc.insert();%>
</body>
</html>
```

Example 11.3

sqlbean.java

```
packageankur;
importjava.io.*;
importjava.sql.connection;
importjava.sql.drivermanager;
importjava.sql.statement;
importjava.sql.resultset;
publicclasssqlbean{
privatestringusername;
privatestringpassword;
 privatestringconnectionurl;
```

```
 privateconnectionconnection=null;
 privateresultsetrs=null;
 privatestatementst=null;
//string connectionurl = "jdbc:mysql://192.168.10.59/sqldata";
public sqlbean()
{
 try{
    // load the database driver
  class.forname("sun.jdbc.odbc.jdbcodbcdriver");
     // get a connection to the database
connection=drivermanager.getconnection("jdbc:odbc:ankur");
}catch(exception e){
system.out.println("exception is ;"+e);
}
 }
   publicvoidsetusername(string username)
  {
 this.username= username;
  }
  publicstringgetusername()
  {
  return(this.username);
  }
  publicvoidsetpassword(stringpassword)
  {
   this.password=password;
  }
   publicstringgetpassword()
  {
  return(this.password);
```

```
 }
 publicvoidinsert()
 {
  try
  {
stringsql="insertintoank(username,password)
     values('"+username+"','"+password+"')";
 statements=connection.createstatement();
 s.executeupdate(sql);
 s.close ();
 }
catch(exception e)
{
 system.out.println("exception is ;"+e);
 }
}
}
```

Example 11.4

How to access database usin JSP:

Check.jsp

```
<%@pageimport="java.sql.*" %>
<%@pageimport="java.lang.object" %>
<%@pageimport="java.io.*" %>
<%@pageimport="java.util.enumeration" %>
<html>
<head>
</head>
% connectionconn=null;
  intf=0;
  try
```

```
   {
class.forname("sun.jdbc.odbc.jdbcodbcdriver");
 stringname=request.getparameter("username");
  stringpass=request.getparameter("password");
 conn=drivermanager.getconnection("jdbc:odbc:law");
statementst=conn.createstatement();
 stringsql="selectusername,passwordfromauth_user";
resultsetrs=st.executequery(sql);
while(rs.next())
 {
  stringname = rs.getstring("username");
  stringpassword = rs.getstring("password");
 if(name.equalsignorecase(name) && pass.equalsignorecase(password))
 {
  f=1;
  break;
 }
else
 {
  f=0;
 }
 }
  if(f==1)
 {
  %> <jsp:forwardpage="csearch1.jsp" />
  < %
  }
  else
  {
  % > notvalidusernameandpassword.tryagain. < %
  }
```

```
}catch(sqlexception e1)
{
 %> notvalidusernameandpassword. tryagain. <%
}
catch(classnotfoundexception e2)
{
%> notvalidusernameandpassword. tryagain. <%
}
catch(exceptione)
{
 %> notvalidusernameandpassword. tryagain. <%
}
if(conn!=null)
{
conn.close();
}
%>
```

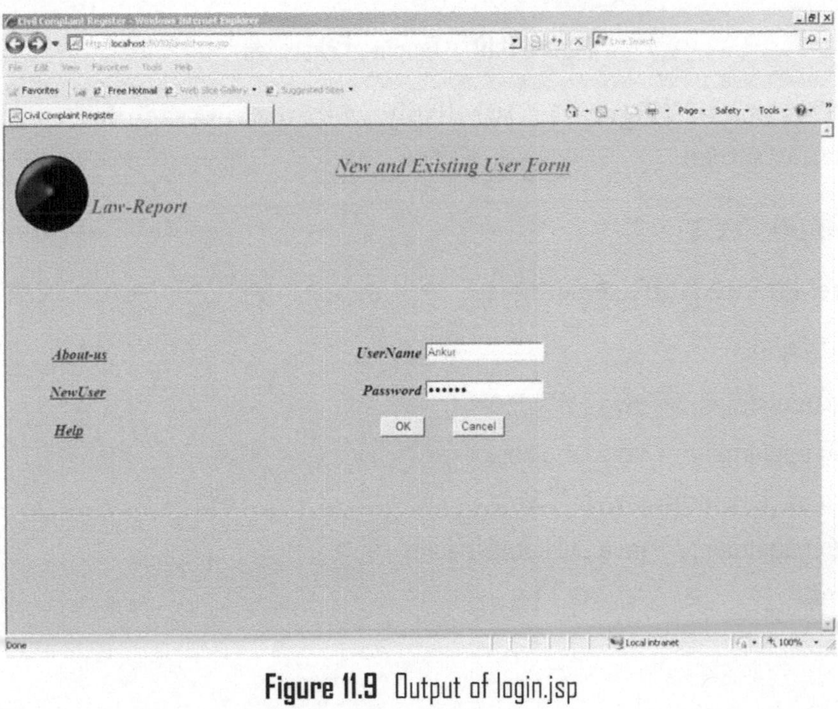

Figure 11.9 Output of login.jsp

This screen (Figure 11.9) display two text box of username, password. Now enter username, password and press ok.

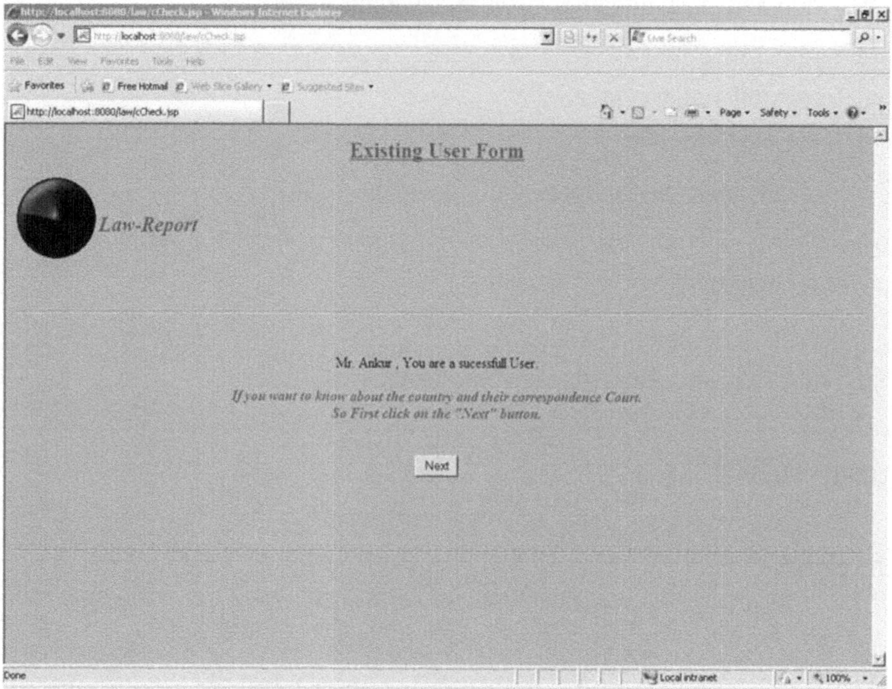

Figure 11.10 Output of check.jsp

This screen (Figure 11.10) disply welcome message for entered user in previous screen.

Example 11.5

Store data in database using jsp:

new1.jsp

```
<%@pageimport="java.sql.*" %>
<%@pageimport="java.lang.object" %>
<%@pageimport="java.io.*" %>
<%@pageimport="java.util.enumeration" %>
<html>
<head>
```

```
<script>
function ba()
{
window.history.back();
}
</script>
</head>
<%
connectionconn=null;
 try
 {
 int psw = 8;
  class.forname("sun.jdbc.odbc.jdbcodbcdriver");
  conn=drivermanager.getconnection("jdbc:odbc:law ","  "," ");
  stringname1=request.getparameter("user");
   stringpass1=request.getparameter("pass");
   statement st=conn.createstatement();
   statement st1=conn.createstatement();
   enumerationparameters=request.getparameternames();
   string ss="";
 resultset rs=st.executequery("select * from auth_user where
       \"password\"='"+pass1+"' ");
while(rs.next())
  {
  stringpass2=rs.getstring("password");
  ss=pass2;
   }
   if(pass1.length()>8)
   {
   %> <a href="cnew1.jsp">notmorethan8char </a> <%
   }
```

```
 else if(pass.equalsignorecase(ss))
 {
% ><a href = "cnew1.jsp">duplicatepasswordisnotallow </a>< %
}
elseif(parameters.hasmoreelements())
 {
  string user1=request.getparameter("user");
  string ps=request.getparameter("pass");
   st1.executeupdate("insert into auth_user(\"username\",\"password\")
     values('"+user1+"','"+ps1+"')");
% ><a href="chome.jsp">sucessfullogin </a>< %
  }
  }catch(sqlexception e1)
{
% ><a href= "cnew1.jsp">notvalidvalue. tryagain </a>< %
}
catch(classnotfoundexception e2)
{
% ><a href= "cnew1.jsp">notvalidvalue. tryagain </a>< %
}
catch(exceptione)
{
 % ><a href = "cnew1.jsp">notvalidvalue. tryagain </a>< %
}
if(conn!=null)
{
 conn.close();
}
% >
<tablealign=centerborder=0cellspacing=0cellpadding=0 width=20%>
<tr>
```

```
<td> <inputtype="button"name=butt1value=backonclick="ba()"> </td>
</tr>
</table>
</body>
</html>
```

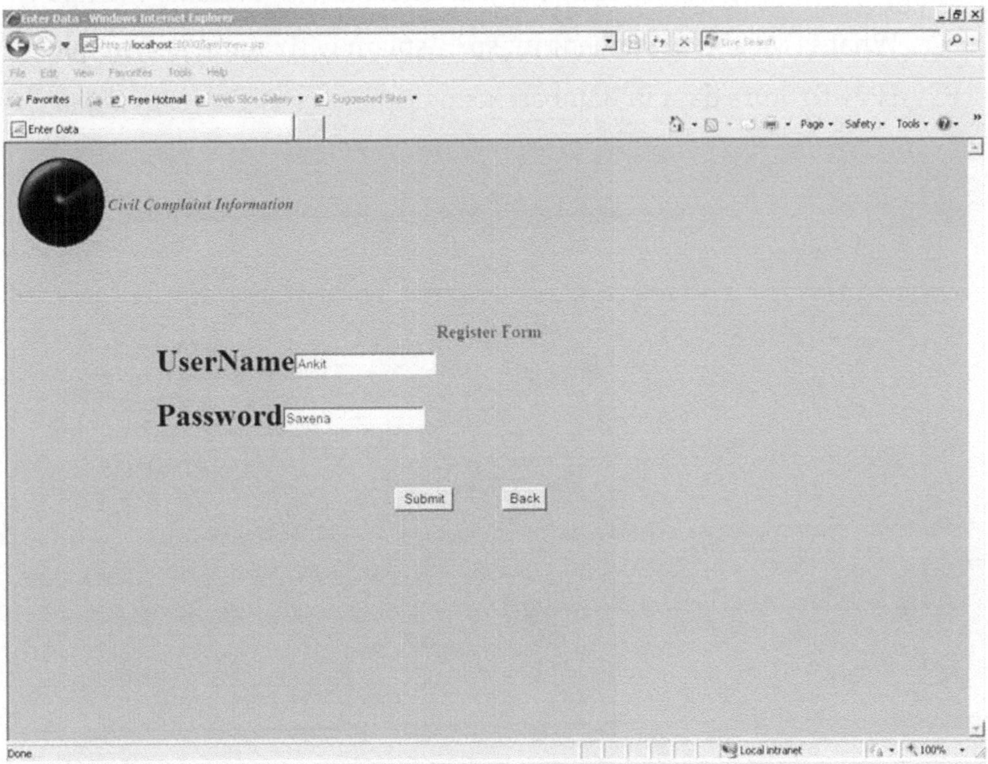

Figure 11.11 output of newl.jsp

This screen (Figure 11.11) display two text boxes which display username, password. Now enter username, password and press submit and this user name, password is stored in database.

REVIEW QUESTIONS

1. What is the role of JDBC in jsp programming?
2. Explain JDBC Architecture with example.
3. What is the basic difference between jdbc type1 and jdbc type2 driver?
4. What is prepare statement in jsp? Explain with suitable example?
5. What is callable statement in jsp? Explain with suitable example?
6. How to Store data in database using jsp?

12
HANDLING COOKIES

12.1 COOKIE

Cookies are the standard mechanism provided by the HTTP protocol for a web server (or a group of web servers sharing the same internet domain)to store small amount of persistent data in a user's web browser for later retrieval. By default, cookie expires as soon as the user exits the browser application. Alternatively, they may be configured to persist across browser sessions until a specified expiration data.

The data stored in a cookie is set by the web server, and therefore can contain only information known to the server. For security reason, a cookie may be retrived only by the server that supplied it. Optionally , a cookie can be made accessible to other servers in the same domain as the originating server. A cookie can also be restricted to a specific URL directory hierarchy on the server or serves from which it is accessible. In addition to the data stored, cookie is assigned a name, a server can then set multiple cookie and distinguish between them via their names.

Cookie is name-value pair, which is stored on the client machine. By default the cookie is implemented in most of the browsers. If we want, then we can also disable the cookie. For security reasons, cookie-based session management uses two types of cookies.

1. *Non-secure session cookie* This cookie can flow between the browser and server under the SSL or non-SSL connection.

2. *Secure authentication cookie* It is used to authenticate the data. This cookie flows over SSL. This type of authentication cookie are used where the information security is very important.

12.2 MANAGING COOKIE

Cookies are set by a web server via HTTP response headers. Whenever a browser requests a URL whose server and directory match those of one or more of its stored cookie, the corresponding cookie are sent back to the server in the form of request headers. If that URL is for JSP page, the page can access those cookies via the getCookie() method of the request implicit object. In a similar manner, cookies are set by a JSP page via the addCookie() method of the response implicit object.

For both methods, HTTP cookies are represented as instance of the javax.servlet.http.Cookie class. The getCookies() method of the request object returns an array of Cookie instance, while the addCookie() method of the response objects takes an instance of the class as its sole argument as shown in Table 12.1.

Table 12.1 HTTP Cookie Method

Implicit objects	Method	Description
Request	getCookies()	Returns an array of the cookie
Response	AddCookies(cookie)	Sends a cookie to the browser for storage or modification.

12.3 Cookie Class

Interacting with the cookie in a JSP page, therefore, is accomplished by manipulating instance of the javax.servelt.http.Cookie class. A single constructor is provided for creating new instance, which takes two String argument, the name of the cookie and corresponding value, as in the following example statement.

```
Cookie cookie =new Cookie("Favorite","chip");
```

Here the first argument represent the name of the cookie and the second its value.

Cookies are named and have a single value. They may have optional attributes, including a comment presented to the user, path and domain qualifiers for which hosts see the cookie, a maximum age, and a version.

The following constructs a cookie with a specified name and value.

```
Cookie(java.lang.String name, java.lang.String value)
```

See the following section on the response object on how to create cookies.

Cookie objects have the following methods as shown in Table 12.2.

Table 12.2 Methods of cookie objects and their descriptions

Method	Description
getComment()	Returns the comment describing the purpose of this cookie, or null if no such comment has been defined.
getMaxAge()	Returns the maximum specified age of the cookie.
getName()	Returns the name of the cookie.
getPath()	Returns the prefix of all URLs for which this cookie is targeted.
getDomain()	Returns the server or domain from which the cookie may be accessed.
getSecure()	Indicate whether the cookie accomplished HTTP or HTTPS requests.
GetValue()	Returns the value of the cookie.
setComment(String)	If a web browser presents this cookie to a user, the cookie's purpose will be described using this comment.
setMaxAge(int)	Sets the maximum age of the cookie. The cookie will expire after that many seconds have passed. Negative values indicate the default behaviour. The cookie is not stored persistently, and will be deleted when the user web browser exits. A zero value causes the cookie to be deleted.
setPath(String)	This cookie should be presented only with requests beginning with this URL.

Table 12.2 (Continues)

Method	Description
SetValue(String)	Sets the value of the cookie. Values with various special characters (white space, brackets and parentheses, the equals sign, comma, double quote, slashes, question marks, the "at" sign, colon, and semicolon) should be avoided. Empty values may not behave the same way on all browsers.
setDomain()	Sets the server or domain from which the cookie may be accessed .
setSecure(name)	Returns the value of a single request header, as an integer.

Example 12.1

Using Cookies Now we will write code in JSP file to set and then display the cookie.

Create Form Here is the code of the form (cookieform.jsp) which prompts the user to enter his/her name

```
<%@ pagelanguage="java" %>
<html>
<head>
<title>CookieInputForm</title>
</head>
<body>
<formmethod="post"action="setcookie.jsp">
<p> <b>Enter Your Name:</b> <inputtype="text"name="username"> <br>
<inputtype="submit"value="OK">
</form>
</body> <html>
```

Above form prompts the user to enter the user name. User input are posted to the setcookie.jsp file, which sets the cookie. Here is the code of setcookie.jsp file.

```
<%@ pagelanguage="java"import="java.util.*"%>
<%
Stringusername=request.getParameter("username");
if(username==null) username="";
Datenow=new Date();
Stringtimestamp= now.toString();
Cookiecookie=newCookie("username",username);
cookie.setMaxAge(365*24*60*60);
response.addCookie(cookie);
%>
<html>
<head>
<title>Cookie Saved</title>
</head>
<body>
<p> <a href="value.jsp">Next  to view the cookie value</a> <p>
</body> <html>
```

Above code sets the cookie and then displays a link to view cookie page. Here is the code of display cookie page (value.jsp).

```
< %@pagelanguage="java" %>
<%
StringcookieName="username";
Cookiecookies[]= request.getCookies ();
CookiemyCookie= null;
if(cookies != null)
{
for(inti= 0; i< cookies.length; i++)
{
if(cookies[i].getName().equals (cookieName))
{
```

```
myCookie=cookies[i];
break;
}
}
}
%>
<html>
<head>
<title>ShowSavedCookie</title>
</head>
<body>
<%
if(myCookie== null) {
%>
NoCookiefoundwiththename<%=cookieName%>
< %
}else{
%>
<p> Welcome:<%=myCookie.getValue()%>.
< %
}
%>
</body>
<html>
```

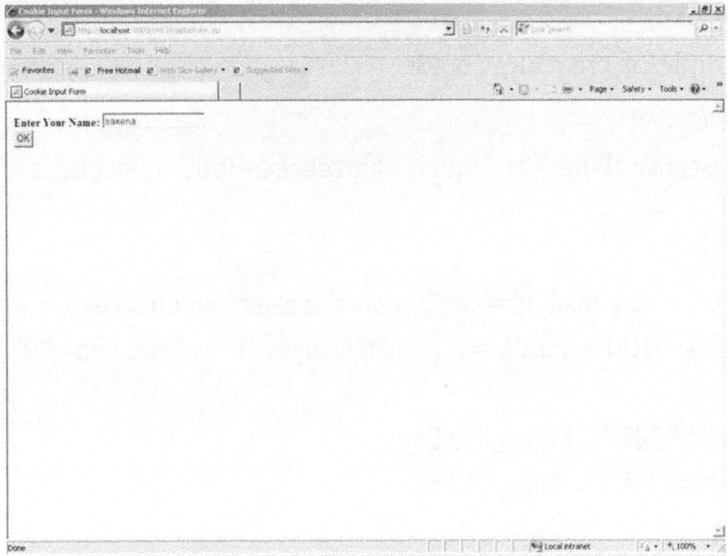

Figure 12.1 Output of cookieform.jsp

This screen (Figure 12.1) display name in textbox. Now press ok.

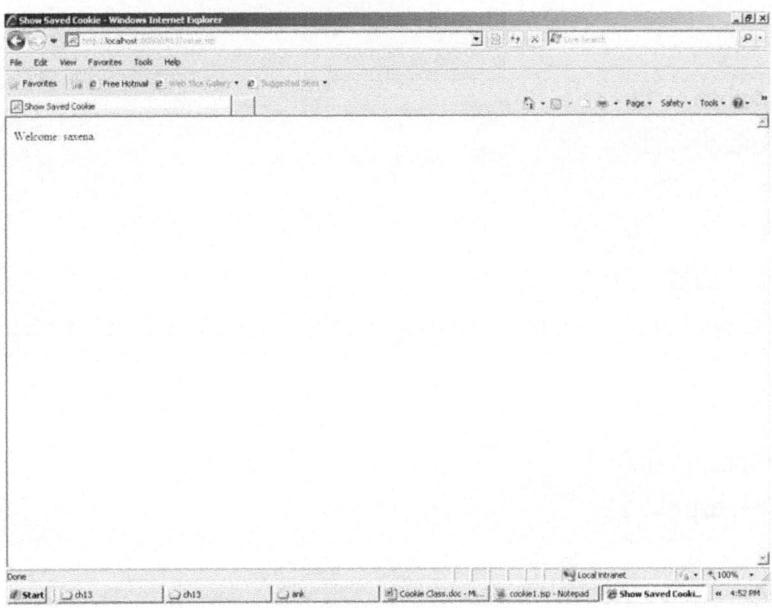

Figure 12.2 Output of value.jsp

This screen (Figure 12.2) display name entered by previous screen using action setcookie.jsp

***Example* 12.2**

JSP Example of creating Cookie

createCookie.jsp

```
<%@pagecontentType="text/html; charset=iso-8859-1"language="java" %>
<html>
<body>
<formname="frm"method="get"action="createNewCookie.jsp">
<tablewidth="100%"border="0"cellspacing="0" cellpadding="0">
 <tr>
 <td width="22%"> </td>
  <td width="78%"> </td>
 </tr>
 <tr>
 <td>Cookie value Set </td>
  <td> <input type="text" name="cookieSet" /> </td>
 </tr>
 <tr>
  <td> </td>
 <td> </td>
  </tr>
  <tr>
  <td> </td>
 <td> <input type="submit" name="submit" value="Submit"> </td>
 </tr>
  <tr>
  <td> </td>
   <td> </td>
  </tr>
</table>
</form>
</body>
</html>
```

Example 12.3

createNewCookie.jsp

```
< %@pagelanguage="java" import="java.util.*"%>
< %
StringcookieSet=request.getParameter("cookieSet");
Cookiecookie=newCookie("cookieSet",cookieSet);
response.addCookie(cookie);
% >
<html>
<head>
<title>Cookie Create Example</title>
</head>
<body>
< %
 Cookie[]cookies=request.getCookies();
  for (int i=0; i<cookies.length; i++) {
out.println(cookies[i].getName()+" : "+cookies[i].getValue()+"<br/>");
 }
 %>
</body>
</html>
```

```
Cookiecookie=newCookie("cookieSet",cookieSet);
```

We are creating new object of cookie class.

```
HerenewCookie("Key","value");
```

Then we are adding in cookie of response object.

`response.addCookie(cookie);` This is adding in cookie object new data.

Expire JSP cookie

```
Cookiecookie=newCookie("cookieSet",cookieSet);
cookie setMaxAge(0);
```

This will expire cookie, with 0 second. We can set cookie age how long is to reside for user.

Retrieve cookie from JSP Cookie is in array object, we need to get in request.getCookies() of array.

```
Cookie[]cookies=request.getCookies();
for (int i=0; i<cookies.length; i++) {
out.println(cookies[i].getName()+" : "+cookies[i].getValue()+"<br/>");
}
```

Example 12.4

Setting and Reading Cookies cookie.jsp

```
<html>
 <head>
  <title>setting and reading cookies</title>
  </head>
  <body
   <%
   cookie[] cookies = request.getcookies();
   boolean foundcookie=false;
   for(int i = 0; i < cookies.length; i++){
  cookie c = cookies[i];
   if (c.getname().equals("color")){
    out.println("bgcolor = "+c.getvalue());
    foundcookie=true;
     }
     }
   if (!foundcookie) {
   cookiec=newcookie("color", "pink");
     c.setmaxage(24*60*60);
    response.addcookie(c);
     }
```

```
    %>
    >
    <h1>setting and reading cookies</h1>
    this page will set its background color using a cookie.
  </body>
</html>
```

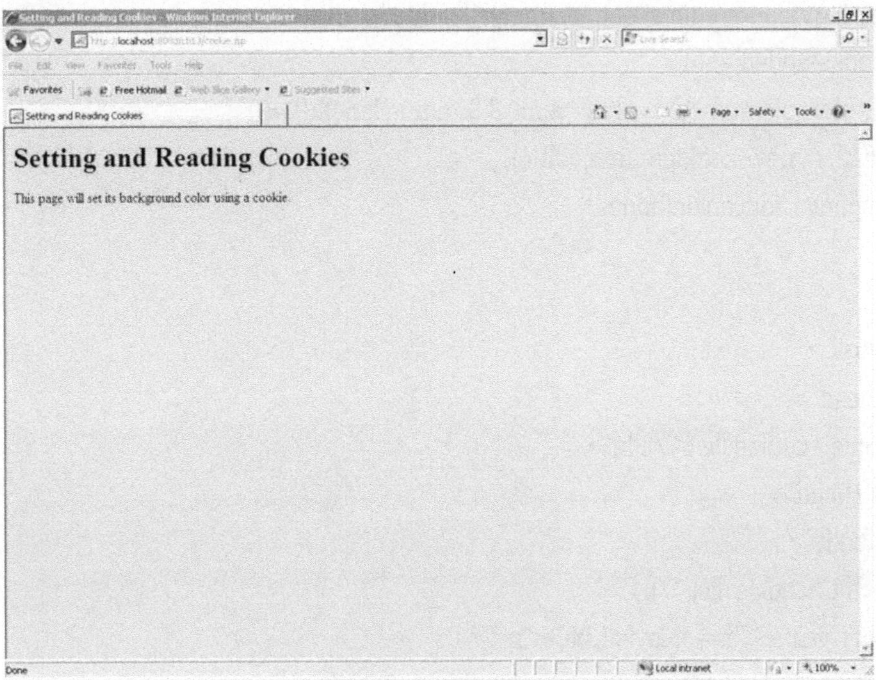

Figure 12.3 Output of cookie.jsp

This screen (Figure 12.3) display background color using cookies.

***Example* 12.5**

cookie1.jsp

```
< %@ page import="java.util.*"%>
< %
  cookie[]cookies=request.getcookies();
  if(cookies==null)
  cookies=newcookie[0];
```

```
  datethisvisit=newdate();
 //adddatacookietosession
  cookiec=newcookie("lastvisit",""+thisvisit);
 response.addcookie(c);
 // add cookie to session
  stringname=request.getparameter("name");
  stringvalue=request.getparameter("value");
 cookieadded=null;
  if (name!=null && value!=null && name.length()>0) {
added = new cookie(name,value);
response.addcookie(added);
 }
%>
<html>
 <head>
 <title>cookie list</title>
  </head>
  <body>
  <h1>cookie list</h1>
  this visit: <%= thisvisit %><br>
  number of cookies: <%= cookies.length %><br>
  <h2>cookies</h2>
  <%
  for (int i=0; i<cookies.length; i++) {
   out.println(cookies[i].getname()+":\t"+
  cookies[i].getvalue()+"<br>");
 // check if added cookie already present
  if (added!=null && added.getname().equals(cookies[i].getname()))
  added= null;
  }
  if (added != null)
```

```
  out.println("new cookie: "+added.getname()+":\t"+
 added.getvalue()+"<br>");
    %>
<h2>new cookie</h2>
 <form>
 <p>name:<input type='text' name='name'></p>
 <p>value:<inputtype='text' name='value'></p>
 <inputtype='submit'value='addnewvalue'>
</form>
</body>
</html>
```

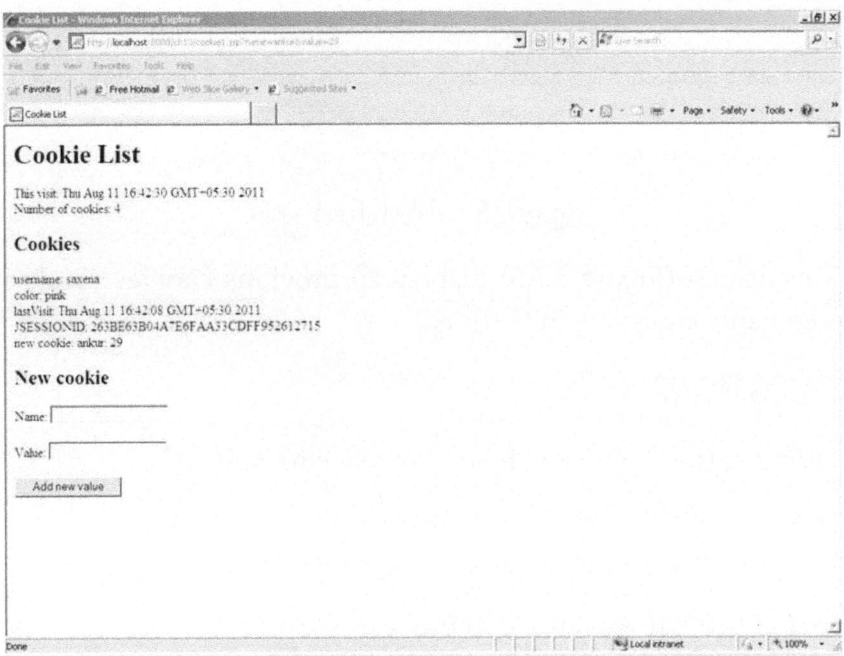

Figure 12.4 Output of cookie1.jsp

This screen (Figure 12.4) display all previous cookies result and display new cookie like name and value in textbox.

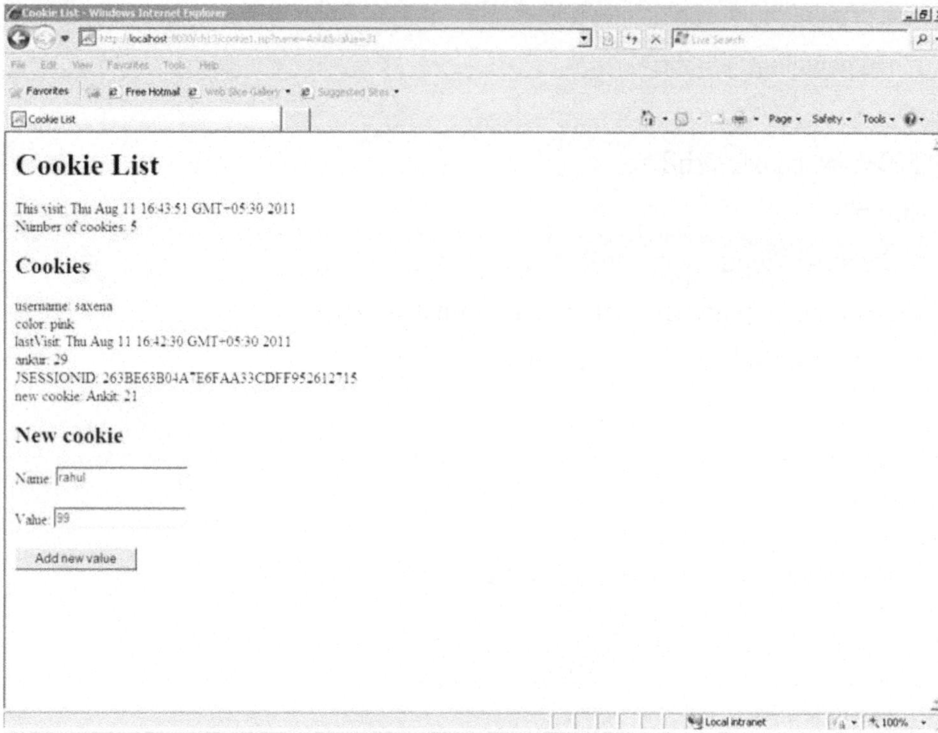

Figure 12.5 Output of cookie1.jsp

This screen (Figure 12.5) display all previous cookies result and display entered name and value in textbox.

REVIEW QUESTIONS

1. What is the cookie and how to manage in jsp?
2. How to create a cookie class in jsp?
3. Explain different method of cookie object?
4. What is the procedure to Expire JSP COOKIE?
5. How to retrieve cookie from JSP?

13

JSP CUSTOM TAGS

13.1 THE NEED FOR A TAG EXTENSION MECHANISM

JSP supports the use of JavaBeans to provide server side components. The tags for using Java Beans are fixed and rather limiting. We can access properties, and we have to use method calls from scriptlets for more involved initialization and so forth.

If we want to apply the Model View Controller (MVC) pattern to web application, with the back-end data providing the model, the business logic acting as the controller, then the JSPs should provide the view of data. It would be big improvement to be able to separate the view and controller.

Web applications often want to provide multiple views on the same data to different channels generating, for example, WML and HTML, or different pages for devices with different abilities, it is important in these applications to separate the view from the controller so that the same controller code can create different views of the same data. Simplifying the page is important too, so that it can be written, customized and tested quickly, by the people whose principle is not the mechanism of the system, but in getting information onto a page in form that's useful and aesthetic.

13.2 WHAT IS A CUSTOM TAG

A tag extension is a java class in fact it is a java bean which implements one of two interface. The details of which we we will look now. The JSP specifications states, the tags have a richer run-time protocol than JavaBean, for these reasons.

- Initialization can involved properties, without which the tag won't run.
- Setting and getting properties doesnot cover all bases there is little control logic available within this bean. We have to control the bean externally, and we could do that with another bean, but that would not be a very general component it would need to be rewritten, compiled and tested for each JSP.
- Beans are not as aware by default of their context in the JSP page as they need to be, there is no default concept of a parent or page Context object within a bean.

13.3 ROLE OF A CUSTOM TAG

A key advantage of JSP over many of the other commonly used dynamic content system is its ability to separate presentation from implementation through the use of HTML like tags. By avoiding the use of JSP elements that embed scripting language code in the page, maintenance of JSP pages is greatly simplified, and the opportunity to reuse the Java Code that provides the underlying functionality is preserved.

Unfortunately, JSP provides only three built in actions for interacting with java-beans objects <jsp:useBean>, <jsp:setProperty>, <jsp:getProperty>.With the exception of these three Javabean tags, the only standard means provided by JSP for accessing arbitrary java code are the scripting elements (scriptlets and expressions). If the need of our application cannot be met via the standard beans tags, then the need to embed java code in the page is unavoidable.

Fortunately, JSP provides custom tag as an extension mechanism for adding new action-tags. As such, custom tags libraries can be written to provide added functionality for a JSP application without having to resort to the use of java code within our JSP pages. It is undesirable to implement dynamic HTML generated via JavaBeans properties, because beans are intended to be stand-alone components that are independent of the type of application within which they are used. Dynamic HTML generation is a fairly uncommon requirement outside the context of JSP, suggesting that Beans, which generate html, may be too closely tied to a specific application.

On the other hand, custom tags are explicitly designed to add functionality to the JSP pages including the dynamic generation of page content such as

HTML. If we need to generate HTML content programmatically via java custom tags are ideal implementation technique. Custom tags can be used to insert text into the page, and also to implement flow of control. Attribute can be specified for custom tags as parameter that influence their behaviour. Custom tags can be empty or have bodies, which contains either nested JSP elements(including other custom tags) or tag specific content to be processed by the tag itself. Custom tag can also interact with each other either by requesting information through the hierarchy of nested tags or by introducing new scripting variables, which may be accessed by subsequent custom tags as well as by the standard JSP scripting elements.

As with all other JSP features, custom tags are implemented via java objects. Developers who are creating their own custom tags do it by creating java classes that produce the desired functionality. As such, custom tags can access to full range of java APIs. For example, custom tags can employ the JDBC classes to make database calls, or use the javamail API to send or receive electronic mails.

Of course, embedding too much functionality into custom tags has its disadvantages. In particular, since custom tags are only meant to be accessed from the JSP pages, any operations built into the implementation of these tags can be used only from JSP. To promote reusability, it is preferable to implement the generic functionality via Javabeans, and use custom tags for controlling presentation and translating between Bean properties and methods and the page markup language.

The implementation, is that there are no limits on the types of behaviour that can be implemented via custom tags. If we have need for a specific computational task to be accomplished on a JSP page, an implementation based on a combination of custom tags and JavaBeans can be developed which maintains a strong separation between presentation and implementation.

13.4 THE JSP CUSTOM TAG API

A Tag is the Java class that is linked to a custom tag and gets invoked every time the JSP container encounters the custom tag. The tag handler is so named because it handles the processing of the tag. To be functional, a tag handler must implement an interface in the javax.servlet.jsp.tagext package or extend one of the classes in the same package. Table 13.1 shows methods in tag interface.

Table 13.1 Description of methods in the tag interface

Method	Description
setPageContext(PageContext pc)	This method is invoked by the JSP runtime, prior to doStartTag, to set the page context.
setParent(Tag parent)	Invoked by the JSP runtime, prior to doStartTag, to pass a tag handler a reference to its parent tag.
GetParent	Returns a tag instance that is the parent of this tag.
DoStartTag	Invoked by the JSP runtime to prompt the tag handler to process the start tag for this instance.
DoEndTag	Invoked by the JSP runtime after returning from doStartTag. The body of the action may or may not have been evaluated, depending on the return value of doStartTag.
Release	Invoked by the JSP runtime to indicate the tag handler to perform any cleanup necessary.

***Example* 13.1**

```
package ankur;
import javax.servlet.jsp.*;
import javax. servlet.jsp.tagext.*;
publicclassBasicTagHandlerimplementsTag{
publicvoidsetParent(Tagt){
 }
publicvoidsetPageContext(PageContextp){
```

```
}
publicvoidrelease(){
}
publicTaggetParent(){
 }
publicintdoStartTag(){
returnEVAL_BODY_INCLUDE;
}
publicintdoEndTag(){
returnEV AL_PAGE;
}
}
```

13.5 THE LIFE CYCLE OF A TAG HANDLER

The life cycle of a tag handler implementing the Tag interface is depicted in Figure 13.1. The first step is to obtain an instance of the appropriate class, either from the tag handler resource pool or, if no instance are currently available, by creating one. Next, various tag handler properties are set. First the handler's set PageContext() method is called by the JSP container to assign it the appropriate PageContext object . Next , the handler setParent() method is called this provides access to the page handler instance, if any within whose body by the current handler appears.

After these properties have been set, the attribute values specified by the tag, if any set. Tag attribute are handled like JavaBean properties, via accessor method defined by the tag handler class. A custom tag attribute named id, for instance, should have corresponding getter and setter methods getId(),setId(), or equivalents specified via an associated Bean Info class. These setter methods are called by the JSP container to assign the attribute value specified in the tag to the corresponding tag handler instance calls these setter methods.

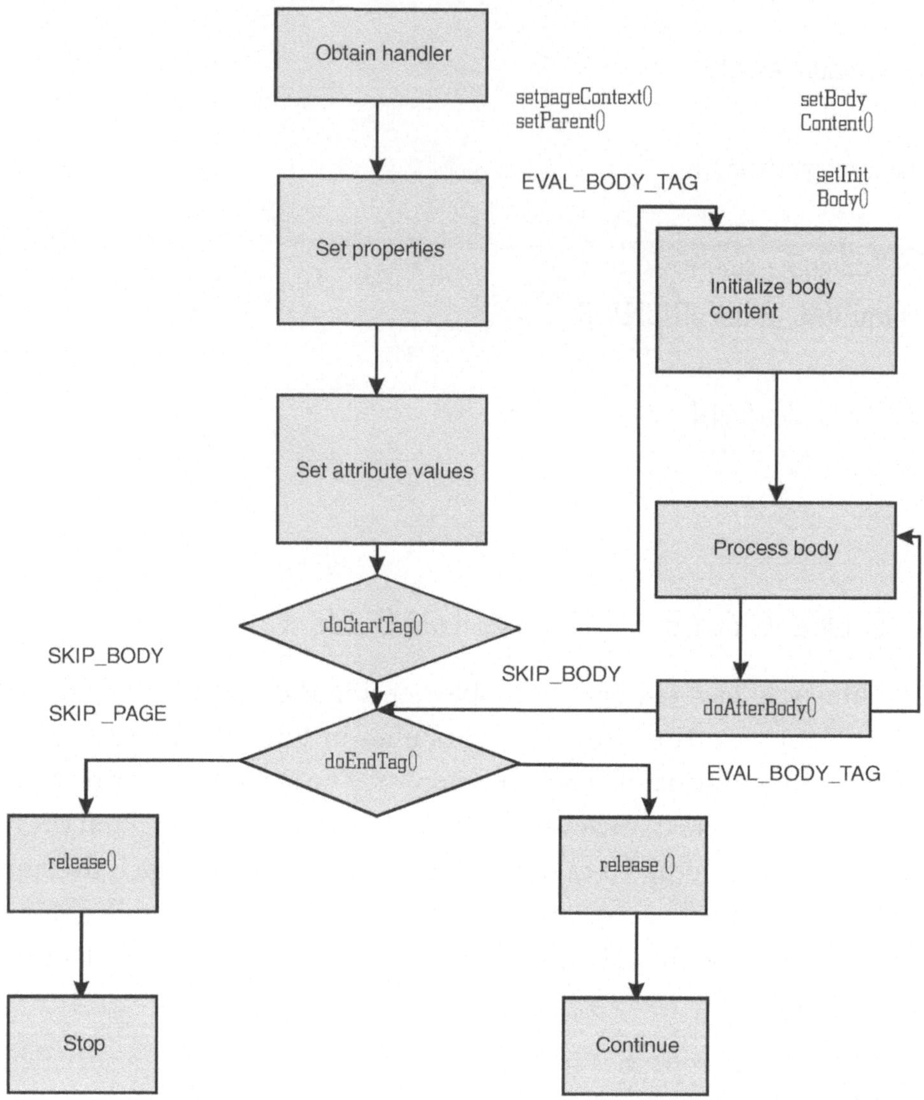

Figure 13.1 Life cycle of handlers implementing tag interface

The JSP container next calls the tag handler's doStartTag() method. At this point, all of the contextual information needed to execute the tag handler will have been provided by the JSP container, so the doStartTag() method can begin performing the action associated with its custom tag. This method should return one of two integer values indicating how processing is to proceed. These two possible values are represented by the class variables Tag SKIP_BODY and Tag

EVAL_BODY_INCLUDES. Tag SKIP_BODY indicates that the body content should be processed normally.

In either case, the next step is to call the handler's doEndTag() method. As with the doStartTag() method, the action performed here are tag specific. Once again the method is expected to return one of two integer value indicating how processing is to proceed, either by Tag SKIP_PAGE or Tag EVAL_PAGE. A return value of Tag SKIP_PAGE from the doEndTag() method indicates to the JSP container that processing of the page should be halted immediately. Any further content on the page, both JSP elements and static text should be ignored and any output generated thus, should be returned to the user's browser. A return value of Tag EVAL_PAGE indicates that page processing should continue normally.

Regardless of the result returned by doEndTag() method, the final processing of a tag handler is for the JSP container to call the handler's release() method. This method gives the tag handler the opportunity cleanup operations—such as resetting its state and releasing any resources it created or obtained while processing the tag, before it is sent to the shared resource pool for subsequent reuse.

Example **13.2**

1. Create a TLD file named taglib.tld as shown below, and save it in the WEB-INF directory.

```
<?xml version="1.0"encoding="ISO-8859-1" ?>
<!DOCTYPE taglib
    PUBLIC "-//Sun Microsystems, Inc.//DTD JSP Tag Library 1.2//EN"
    "http://java.sun.com/dtd/web-jsptaglibrary_1_2.dtd">
    <taglib>
    <tlibversion>2.0</tlibversion>
    <shortname> </shortname>
    <tag>
    <name>myTag</name>
    <tagclass>ankur.MyCustomTag</tagclass>
</tag>
    </taglib>
```

2. Write, compile, and deploy the Java class called MyCustomTag.java given below. Make sure that the class file is located in the directory, under WEB-INF/classes/ankur directory.

```
package ankur;
import javax.servlet.jsp.*;
importjavax.servlet.jsp.tagext.*;
public class MyCustom Tagextends TagSupport{
public int doEndTag() throwsJspException{
 JspWriterout=pageContext.getOut();
 try{
 out.println("HelloAnkurSaxena.");
}
catch(Exception e){
}
return super. doEndTag();
}
}
```

3. Create a JSP file from the code given below , Call it simple.jsp

```
<%@taglib uri="/myTLD" prefix="easy"%>
<easy:myTag/>
```

4. Edit your deployment descriptor(web.xml)file. To use custom tag,you must specify a <taglib> element in your web.xml file. The <taglib > elements must appear after the <servlet> and <servlet-mapping>

```
<!doctype web-app
  public "-//sun microsystems, inc.//dtd web application 2.3//en"
  "http://java.sun.com/dtd/web-app_2_3.dtd">
<web-app>
 <display-name>template</display-name>
 <taglib>
       <taglib-uri>/mytld</taglib-uri>
       <taglib-location>/web-inf/taglib.tld</taglib-location>
```

```
</taglib>
</web-app>
```

***Example* 13.3**

The following example illustrates a tag handler that provides implementation for the setPageContext and doStartTag method.

```
package ankur;
import javax.servlet. jsp.*;
import javax. servlet. jsp. tagext.*;
public class MyCustomTagimplements Tag{
  pagecontext pagecontext;
  public void setparent(Tag t){
  }
public void setpagecontext(pagecontext p){
      pagecontext = p;
}
public void release(){
}
public tag getparent(){
 return null;
 }
public int dostarttag(){
      try{
      jspwriter out=pagecontext.getout();
      out.println("helloankursaxena.");
}
catch(exception e){
}
returneval_body_include;
}
public int doendtag()throws jspexception{
```

```
        returneval_page;
    }
    }
```

The JSP page that contain a custom tag that uses the content substitution TagHandler

```
<%@taglib uri="/mytld" prefix="easy"%>
<easy:mytag/>
```

to test the tag handler, you need jsp page given below.

```
<?xml version="1.0"encoding="iso-8859-1" ?>
<!doctype taglib
  public "-//sun microsystems, inc.//dtd jsp Tag library1.1//en"
"http://java.sun.com/j2ee/dtds/web-jsptaglibrary_1_1.dtd">
<taglib>
      <tlibversion>1.0</tlibversion>
      <shortname></shortname>
      <tag>
      <name>mytag</name>
      <tagclass>ankur.mycustomtag</tagclass>
</tag>
</taglib>
```

Example 13.4

Using attribute in a custom tag The following example presents a tag handler named doubtag that doubles an integer and outputs the results to the web browser.

```
package ankur;
import javax.servlet.jsp.*;
import javax.servlet.jsp.tagext.*;
public classdoublerTag implements Tag{
 private int number;
 public void setNumber(int number){
```

```
        this.number=number;
    }
    pagecontext pagecontext;
    public void setParent(tag t){
    }
   public void setPageContext(pagecontext p){
        pagecontext= p;
   }
  public void release(){
        }
  public Tag getparent(){
        returnnull;
  }
  public int dostarttag(){
        try{
        jspwriter out=pagecontext.getout();
        out.println("doubleof "+number+"is"+(2*number));
  }
  catch(exception e) {
  }
  return Eval_Body_Include;
        }
  public int doendtag()throws jspexception {
        returneval_page;
  }
        }
```

Tld file

```
<%@taglib uri="/mytld"prefix="easy"%>
<easy:mytag number="12"/>
```

jsp file(11.12)

```
<?xml version="1.0"encoding="iso-8859-1"?>
```

```
<!doctype taglib
     public "-//sun microsystems, inc.//dtd jsp tag library1.1//en"
     "http://java.sun.com/j2ee/dtds/web-jsptaglibrary_1_1.dtd">
<taglib>
 <tlibversion>2.0</tlibversion>
 <shortname> </shortname>
  <tag>
<name>my Tag </name>
     <tagclass>ankur.doublertag</tagclass>
 <attribute>
     <name>number</name>
     <required>true</required>
</attribute>
</tag>
</taglib>
```

***Example* 13.5**

A power tag The following example demonstrates a tag handler that implements the IterationTag interface and illustrates the use of the doAfterBody method.

The code of the tag handler is given below.

```
package ankur;
import javax.servlet.jsp.*;
import javax.servlet.jsp.tagext.*;
public class PowerTag implements IterationTag{
PageContext pageContext;
private int number;
private int power;
private int counter;
private int result=1;
 //the setter for number
```

```
public void setNumber(intnumber){
    this.number=number;
}
//the setter for power
public void setPower(intpower){
    this.power=power;
 }
 public void setParent(Tagt){
 }
 public void setPageContext(PageContext p){
    pageContext = p;
 }
 public void release(){
 }
 public Tag getParent(){
    returnnull;
 }
 public int doStartTag(){
 return Eval_Body_I    ;
 }
 public int doAfterBody(){
    counter++;
    result*=number;
    if(counter= =power)
    returnSkip_Body;
 else
    returnEval_Body_Again;
 }
 public int doEndTag()throwsJspException{
    System.out.println("doEndTag");
  try{
```

```
  JspWriterout = pageContext.getOut();
  out.println(number+"^"+power+"="+result);
  }
  catch(Exception e){
 }
     return Eval_Page;
}
}
```

TLD FILE

```
<%@ taglib uri="/myTLD"prefix="easy"%>
<easy:myTagnumber="2"power="3"/>
<?xml version="1.0"encoding="ISO-8859-1"?>
<!Doctype taglib
        PUBLIC "-//SunMicrosystems,Inc.//DTD JSP Tag Library 1.1//EN"
        "http://java.sun.com/j2ee/dtds/web-jsptaglibrary_1_1.dtd">
<taglib>
  <tlibversion>2.0</tlibversion>
  <shortname> </shortname>
  <tag>
     <name>myTag</name>
     <tagclass>ankur.PowerTag</tagclass>
      <attribute>
     <name>number</name>
     <required>true</required>
 </attribute>
 <attribute>
     <name>power</name>
     <required>true</required>
</attribute>
</tag>
</taglib>
```

REVIEW QUESTIONS

1. What is custom tag mechanism in jsp?
2. What is JSP Custom Tag API?
3. Explain the Life Cycle of a Tag Handler in jsp?
4. How to use attribute in a custom tag?

14
INTRODUCTION OF MVC

14.1 INTRODUCTION

A design pattern describes a proven solution to a recurring design problem, placing particular emphasis on the context and requirement surrounding the problem, and the consequences and impact of solution.

There are many good reasons to use design pattern.

1. They are proven. You tap the experience, knowledge and insights of developers who have used these pattern successfully in their work.
2. They are reusable. When a problem recurs, you don't have to invent a new solution, you follow the pattern and adapt it, as necessary.
3. They are expressive. Design pattern provide a common vocabulary of solutions, which you can use to express larger solution succinctly.

It is important to remember, that design pattern do not guarantee success. You can only determine whether a pattern is applicable by carefully reading its description, and only after you have applied it in your own work you can determine whether it has helped any. One of these patterns is Model-View-Controller (MVC). The MVC design idiom has become common place, especially in object-oriented systems.

14.2 MVC ACHITECTURE

The goal of the MVC design pattern is to separate the application object(model) from the way it is represented to the user(view), from the way in which the user controls it.

14.2.1 Model

A model represents an application's data and contains the logic for accessing and manipulating that data. Any data that is part of the persistent state of the application should reside in the model objects. The services that a model exposes must be generic enough to support a variety of clients. By glancing at the model's public method list, it should be easy to understand how to control the model's behaviour. A model groups related data and operations for providing a specific service. These group of operations wrap and abstract the functionality of the business process being modelled. A model's interface exposes methods for accessing and updating the state of the model and for executing complex processes encapsulated inside the model. Model services are accessed by the controller for either querying or effecting a change in the model state. The model notifies the view when a state change occurs in the model.

14.2.2 View

The view represents the presentation of the application. The view object refers to the model. It uses the query methods of the model to obtain the contents and renders it. The view is not dependent on the application logic. It remains same if there is any modification in the business logic. In other words, we can say that it is the responsibility of the views to maintain the consistency in its presentation when the model changes.

14.2.3 Controller

Whenever the user sends a request for something then it always goes through the controller. The controller is responsible for intercepting the requests from view and passes it to the model for the appropriate action. After the action has been taken on the data, the controller is responsible for directing the appropriate view to the user. In GUIs, the views and the controllers often work very closely together.

14.3 BENEFITS OF MVC ARCHITECTURE

The MVC architecture has the following benefits.

1. *Multiple views using the same model* The separation of model and view allow multiple views to use the same enterprise model. Consequently, an enterprise's application model components are easier to implement, test and maintain, since all access to the model goes through these components.

2. *Easier support for new type of clients* To support a new type of client, you simply write a view and controller for it and write them into the existing enterprise model.

3. *Clarity of design* By glancing at the model's public method list, it should be easy to understand how to control the model's behaviour. When designing the application, this trait makes the entire program easier to implement and maintain.

4. *Efficient modularity* This design allows any of the components to be swapped in and out as the user or programmer's desire. Even the model changes to one aspect of the program are not coupled to other aspects, eliminating many nasty debugging situations. Also, development of the various components can progress in parallel, once the interface between the components is clearly defined.

5. *Ease of growth* Controllers and view can grow as the model grows and older version of the view and controller can still be used as long as a common interface is maintained.

6. *Distributable* With a couple of proxies one can easily distribute any MVC application by only altering the startup methods of the application.

14.4 MODEL 1 ARCHITECTURE

In model 1 architecture, (Figure 14.1) the application is page-centric. The client browser navigates through a series of JSP pages in which any JSP page can employ a JavaBean that performs business operations. However, the highlight of this architecture is that each JSP page processes its own input. Application implementing this architecture normally have series of JSP pages, where the user is expected to proceed from the first page to the next. If needed, a servlet or an HTML page can be substituted for the JSP page in the series.

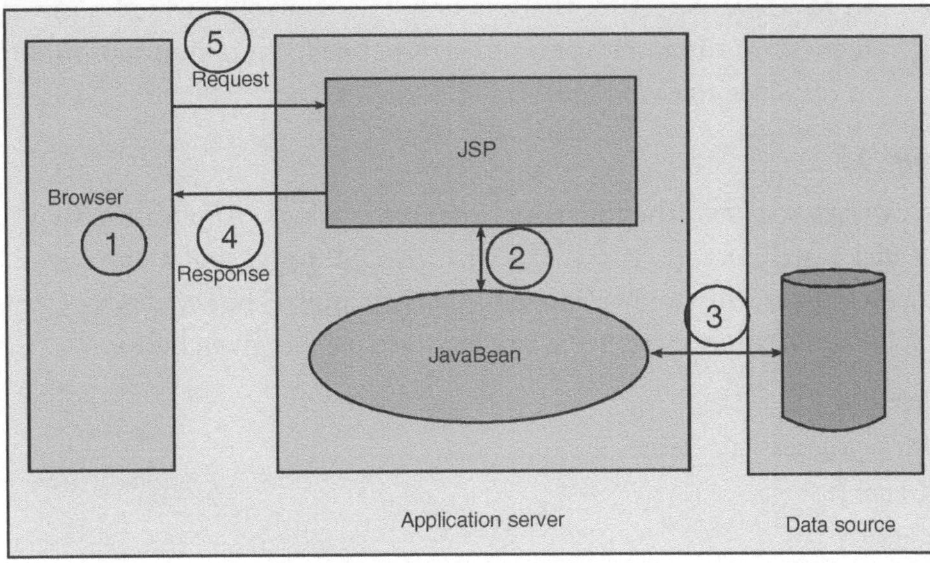

Figure 14.1 JSP model I architecture

In this Architecture (Figure 14.1) browser send a request to jsp, jsp check setter and getter method inside JavaBean. Bean map the data from database and finally jsp send a response to browser.

14.5 DIFFERENCE BETWEEN MODEL 1 AND MODEL 2 ARCHITECTURE

Features of MVC1

1. HTML or jsp files are used to code the presentation. To retrieve the data JavaBean can be used.
2. In MVC1 archictecture all the view, control elements are implemented using Servlets or JSP.
3. In MVC1 there is tight coupling between page and model, as data access is usually done using Custom tag or through Java Bean call.

14.5.1 Features of MVC2

1. The MVC2 architecture removes the page-centric property of MVC1 architecture by separating presentation, control logic and the application state.

2. In MVC2 architecture there is only one controller which receives all the request for the application and is responsible for taking appropriate action in response to each request.

Example 14.1

a login application The following example is a login Application that uses the Model 1 architecture. It consists of three JSP pages and a Java Bean. As the name implies, this application is for user to login. The sequence of this series of JSP application starts from the login.jsp page is given below.

```
login.jsp
<%@pagesession="false" %>
< html >
<head>
<title>login page</title>
</head>
<body>
<formmethod="post"action="process.jsp">
< %
  if (request.getparameter("error")!=null){
% >
loginfailed.pleasetryagain
<br> <hr>
< %
  }
%>
<table>
< tr >
  <td>username</td>
  <td> <inputtype=textname="username"> </td>
</tr>
<tr>
  <td> password</td>
```

```
  <td> <inputtype=passwordname="password"> </td>
</tr>
<tr>
  <tdcolspan="2"> <inputtype=submitvalue="login"> </td>
</tr>
</table>
</form>
</body>
</html>
```

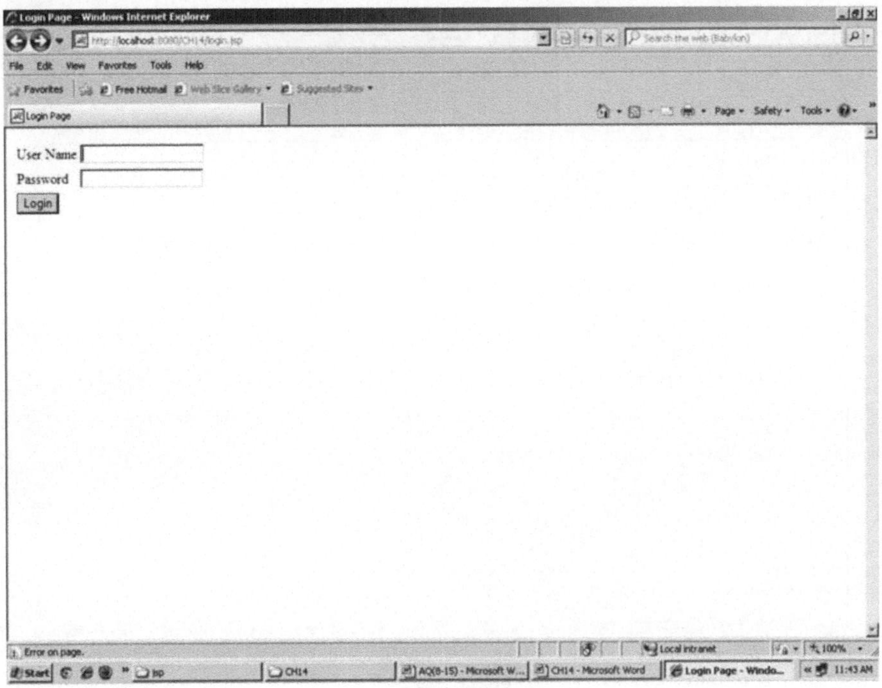

Figure 14.2

This screen (Figure 14.2) display login information. Now enter correct username and password.

The login.jsp page is nothing more than an HTML page with a form. The only code present in the scriptlet after the <FORM> tag that checks to see if the request object contains a parameter called "error", if so, a message is displayed.

```
<%
```

```
 if (request.getParameter("error")!=null){
%>
Loginfailed.Pleasetryagain
<BR> <HR>
<%
 }
%>
```

The Form in the Login page is submitted to the second page process.jsp. which is given below.

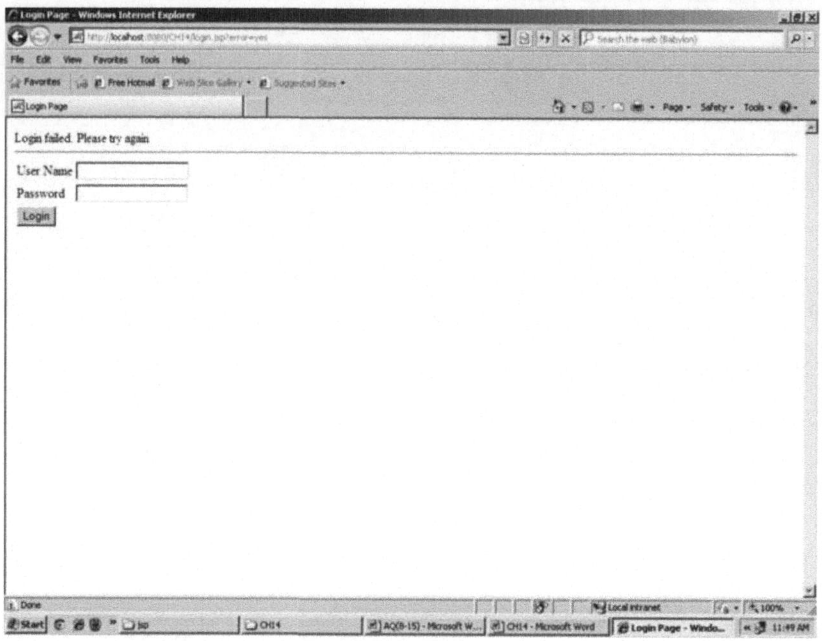

Figure 14.3

This screen (Figure 14.3) is displayed when your username and password are incorrect.

process.jsp

```
<%@page session="false" %>
<jsp:useBeanid="loginBean"scope="page"class="ankur.loginBean"/>
<%
```

```
if(loginBean.login(request.getParameter("userName"),
request.getParameter("password")))
request.getRequestDispatcher("welcome.jsp").forward(request, response);
else
//wehavetousesendRedirectbecausewewanttosendtheerrorpart
   //totheLogin.jsp page.
   //withRequestDispatcher.forward(),theURLwillstillbe
   //thecurrentURL
   response.sendRedirect("login.jsp?error=yes");
%>
```

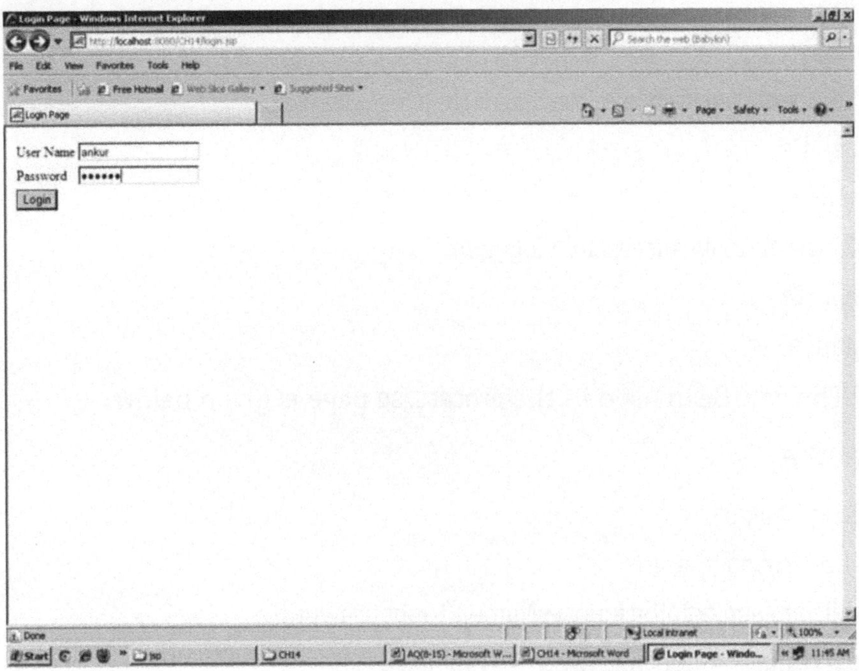

Figure 14.4

This page employs a JavaBean called ankur.LoginBean. The process.jsp page calls the login method of the bean, passing the username parameter and the password parameter of the request object. The login method returns true if the user is an authorized user. It returns false otherwise.

In a successful login the process.jsp page displays the welcome page.

```
if(loginBean.login(request.getParameter("userName"),
 request.getParameter("password")))
request.getRequestDispatcher("Welcome.jsp").forward(request, response);
      .
      .
      .
```

The Welcome.jsp page is given below.

welcome.jsp

```
< %@pagesession="false" %>
<html>
<head>
<title> welcome</title>
</head>
<body>
Welcome.Youhavesuccessfullyloggedin.
</body>
</html>
```

The JavaBean used in the process.jsp page is given below.

loginBean.java

```
package ankur;
publicclassLoginBean{
 publicbooleanlogin(StringuserName,Stringpassword){
if (userName==null||password==null||
 !(userName.equals("ankur") && password.equals("saxena")))
returnfalse;
 else
 returntrue;
}
}
```

In the login method in the JavaBean, successful login occurs if the userName equals "ankur" and the password is same as the "saxena". The Login method is purposely very simple .

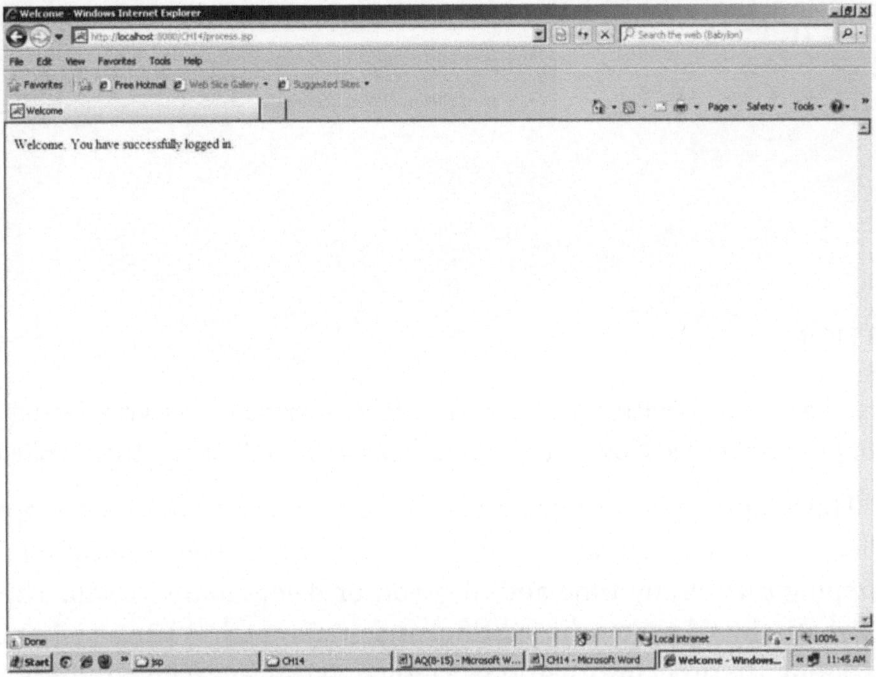

Figure 14.5

This screen (Figure 14.5) display welcome message after login successful.

REVIEW QUESTIONS

1. Explain some good reasons to use design pattern.
2. What is model view controller?
3. Explain Benefits of MVC Architecture.
4. What is the basic difference between Model 1 and Model 2 architecture?

15
JSP PROJECT

DESCRIPTION

This chapter deals with real application of java server pages with the help of online shopping of Books. Now you can purchase any book using this application.

The Shopping Cart program allows visitors to your on-line shopping site to collect books in a virtual shopping cart. They may view the contents of their shopping cart at any time and may add or delete items at will. The program automatically calculates the subtotal and grand total. When a visitor decides to check-out, the order information including the buyer's name, address and billing instruction is e-mailed to your order department (or whomever you choose) and a receipt is sent to the shopper.

Features

- Purchase one or more books.
- Sends receipt to customer.
- Search book for different categories.
- Modify book with different categories.
- Tracks purchases even if user clicks the back button.
- Tracks each customer by Shopper ID.

Application

Before using this application, the user has to register himself. The books could be scarched under four different categories. They are:

1. Search by course
2. Search by book title
3. Search by author
4. Search by publishers

Select the book with different category and place order if you need to purchase more than one book. Then again search the book and place order. After this, the total price of the books display on your screen. Now enter your card details. Your books are deliver within 48 working hours.

Vendor of this project has all rights regarding the book like:

1. Insert new books
2. Modify book
3. Insert new category
4. Modify category

Database: Ebook

Table:Login1

Table 15.1

Field	Data type
name	Varchar(10)
password	Varchar(10)

Table :Book_detailed

Table 15.2

Field	Data type
bookid	Integer
categoryid	Integer
title	Varchar(10)
author	Varchar(10)
publisher	Varchar(10)
edition	Varchar(10)
price	Integer
quantity	Integer
description	Varchar(10)

Table :Category_details

Table 15.3

Field	Data type
categoryid	integer
categoryname	Varchar(10)

Table :Order_detail

Table 15.4

Field	Data type
Ordered	integer
bookid	integer
quantity	Varchar(10)

Table :Order

Table 15.5

Field	Data type
orderid	integer
userid	integer
totalamount	integer
orderdate	date

Table :Temp_detail

Table 15.6

Field	Data type
bookid	integer
categoryid	integer
title	Varchar(10)
author	Varchar(10)
publisher	Varchar(10)
edition	Varchar(10)
price	integer
quantity	integer
description	Varchar(10

Table :User_auth

Table 15.7

Field	Data type
username	Varchar(10)
password	Varchar(10)

Table :User_profile

Table 15.8

Field	Data type
username	Varchar(10)
password	Varchar(10)
firstname	Varchar(10)
middlename	Varchar(10)
lastname	Varchar(10)
Address1	Varchar(10)
Address2	Varchar(10)
city	Varchar(10)
state	Varchar(10)
pincode	integer

login.htm

```
<html>
<head>
<title>Login</title>
</head>
<bodybgcolor="#CCFFFF">
<Center>
<H1> <u> OnlineShoppingCart</u> </H1>
</Center>
<Scriptlanguage="JavaScript">
 function validate()
 {
      if(document.frm.UserName.value==""||document.frm.Password.value=="")
 {
 alert("UserName or Password cannot be blank!!");
 return;
```

```
  }
   document.frm.action="Validate.jsp";
   document.frm.submit();
  }
</SCRIPT>
<formname=frmmethod="POST">
<p align ="center"> <font size ="5"> <b> Login Form </b> </font> </p>
<Table Align="center" height="57">
<TR>
<TD >
     <b>UserName:</b>
</TD>
<TD >
<inputtype ="text"name="UserName"size="20"tabindex="1">
</TD>
</TR>
<TR>
<TD >
     <b>Password:</b>
</TD>
<TD >
<input type="password" name ="Password" size ="20" tabindex ="2">
</TD>
<TRalign="center">
<TDcolspan=2>
<inputtype="button"value ="Login"name ="B1"onclick="validate();" tabindex
     ="3">
</TD>
</TR>
<TR >
<TD > <font size ="4">New User? </font> </TD>
```

```
<TD> <a href ="NewUser.htm"> <font size="4">Register</font> </a> </TD>
</TR>
</Table>
</form>
</body>
</html>
```

Validate.jsp

```
<html>
<body>
<%@pagelanguage="java"import="java.sql.*"%>
<%
Connection conn;
conn=null;
ResultSet rs;
rs=null;
try
{
    StringstrName=request.getParameter("UserName");
    StringstrPass=request.getParameter("Password");
    Class.forName("sun.jdbc.odbc.JdbcOdbcDriver");
    conn=DriverManager.getConnection("jdbc:odbc:Data","scott","tiger");
    PreparedStatement stat= conn.prepareStatement("SELECT * FROM user_auth
    where username=? and password =?");
    stat.setString(1,strName);
    stat.setString(2,strPass);
    rs = stat.executeQuery();
    if(rs.next())
{
    Stringstr1,str2;
    str1=rs.getString(1);
    str2=rs.getString(2);
```

```
        out.println(str1+str2);
        session.putValue("user",strName);
        if(str1.equals("Administrator"))
        {
              response.sendRedirect("Admin.htm");
        }
        else
        {
              out.println("Logged IN");
              response.sendRedirect("SearchCriteria.jsp");
        }
  }
  else
  {
        response.sendRedirect("InvalidUser.htm");
        }
  }
  catch(Exception E)
  {
        out.println("Error "+E);
  }
  finally
  {
        rs.close();
        conn.close();
        }
        %>
  </body>
  </html>
```

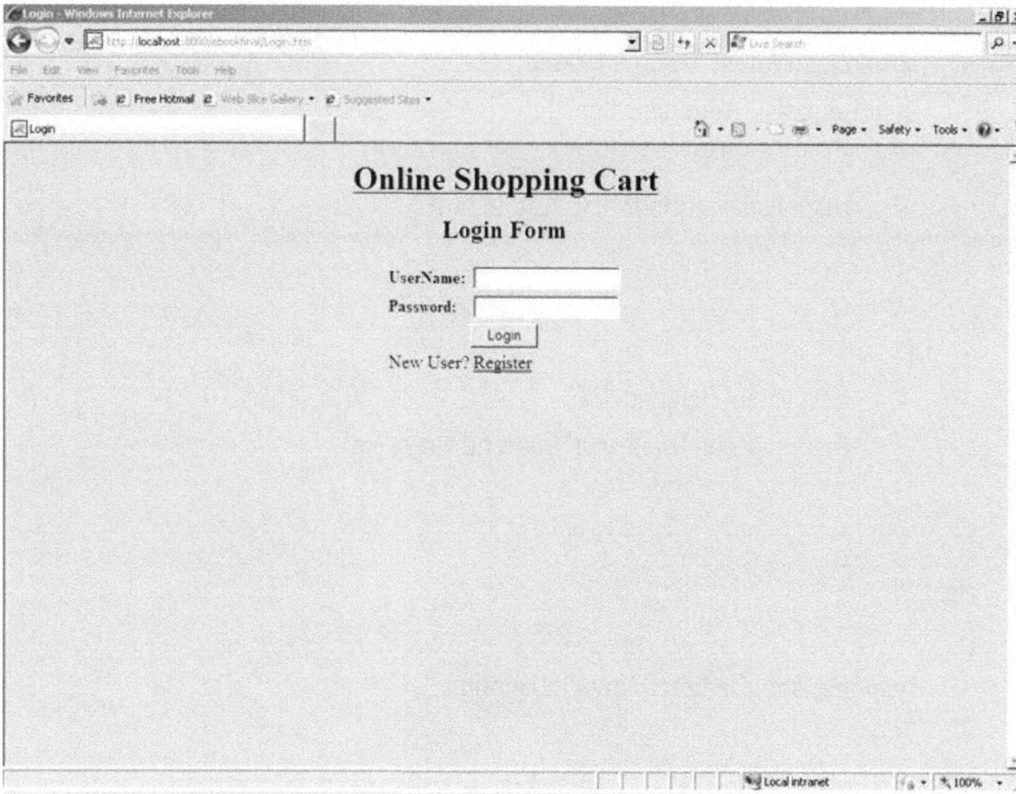

Figure 15.1 Output of Login.htm

This screen (Figure 15.1) show the login page of the project. Now enter username, password.

InvalidUser.htm

```
<html>
<head>
<title>invalid user</title>
</head>
<body bgcolor="#ccffff">
<center>
<h1> online shopping cart </h1>
</center>
<br> <hr>
```

```
<formmethod="post" action="">
<palign ="center">
<fontsize ="5"color =#000000>thespecifieduserdoesnotexist.
</font>
<br>
<br>
<br>
<table align="center" border=0 height="102" width="359">
<tr>
<td height="37" width="190">
    <font size="4">create new account</font>
</td>
<td height="37" width="155">
<p align="left"> <a href="newuser.htm"> <font size="4">click here</font> </
    a> </p>
</td>
</tr>
<tr>
<td height="1" width="190">
<font size="4">login again</font>
</td>
<td height="1" width="155">
<font color="#0000ff"> <a href="login.htm"> <font size="4">click here</
    font> </a> </font>
</td>
</tr>
</table>
</form>
</body>
</html>
```

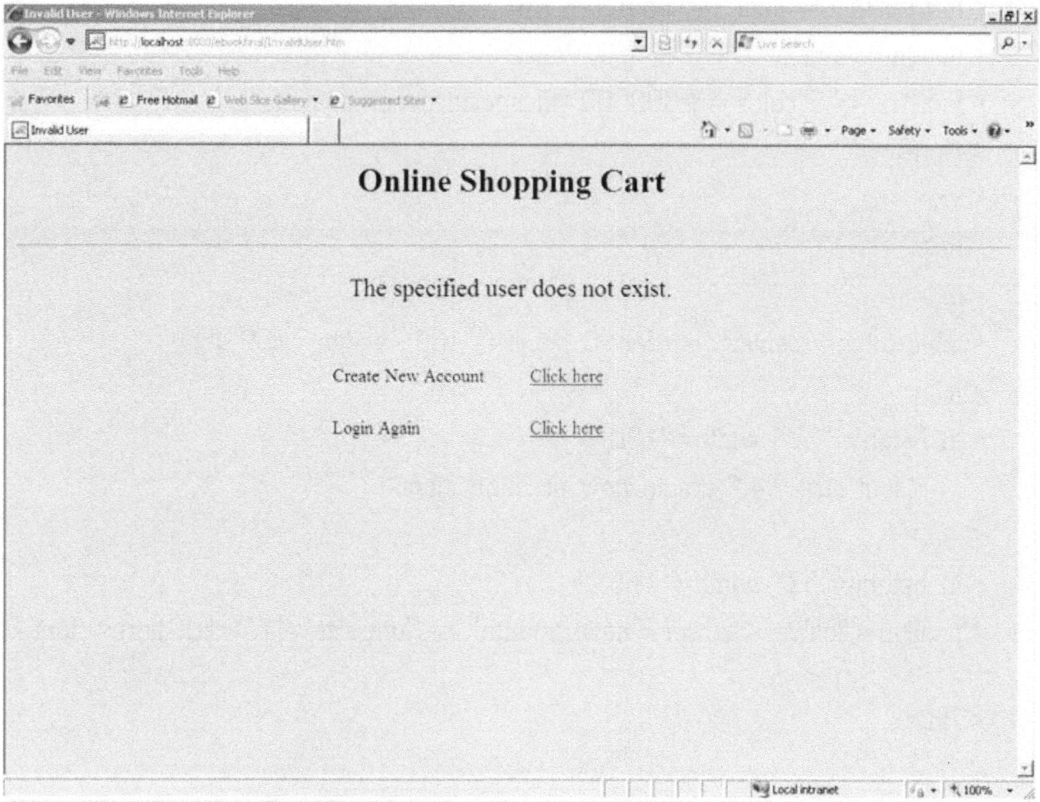

Figure 15.2 Output of InvalidUser.htm

This screen (Figure 15.2) show two links; first to create new account and second to existing user.

NewUser.htm

```
<html>
<head>
<title> NewUserRegistration</title>
</head>
<bodybgcolor="#CCFFFF">
<Center>
<H1> <u>OnlineShoppingCart</u> </H1>
</CENTER>
```

```
<formname="form1"method="POST" >
 <Scriptlanguage="JavaScript">
  functionvalidate()
  {
  if(document.form1.UserName.value= ="")
  {
  alert("UserNamecannotbeblank");
  return;
  }
  if(document.form1.Password.value= ="")
  {
  alert("Password cannot be blank");
  return;
  }
  if(document.form1.Password.value!= document.form1.Password2.value)
  {
  alert("The two passwords do not match");
  return;
  }
  if(document.form1.FirstName.value= ="")
  {
  alert("First Name cannot be blank");
  return;
  }
  if(document.form1.LastName.value= ="")
  {
  alert("Last Name cannot be blank");
  return;
  }
 if( document.form1.Address1.value= ="")
 {
```

```
alert("Addresscannotbeblank");
return;
}
if( document.form1.City.value=="")
{
alert("Citycannotbeblank");
return;
}
if( document.form1.State.value=="SelectState")
{
alert("Youmustselectastate");
return;
}
if(document.form1.PinCode.value=="")
{
alert("PinCodecannotbeblank");
return;
}
if(document.form1.Email.value=="")
{
alert("EmailIDcannotbeblank");
return;
}
if(document.form1.Phone.value=="")
{
alert("PhoneNumbercannotbeblank");
return;
}
document.forms[0].action="Sign.jsp";
document.forms[0].submit();
}
```

```
</SCRIPT>
<p align="center">
<b> <font size="5"> New User Registration</font> </b> 
        
</p>
<p align="left">
 <b> <font color="#FF0000">( Field marked * are necessary)</font> 
               
 </p>
 <div align="left">
     <table border="0" width="100%">
      <tr>
      <td width="23%"> <b>User *:</b> </td>
      <td width="77%">
      <b>
      <input type="text" name="UserName" size="20" tabindex="1">
      </b>
 </td>
 </tr>
 <tr>
     <td width="23%">
      <b>
 Password*:</b> </td>
 <td width="77%"> <b> <input type="password" name="Password" size="20"
     tabindex="2">
 </b>
     </td>
</tr>
<tr>
     <td width="23%">
            <b>
            Confirm Password*:</b>
```

```
</td>
<td width="77%"> <b> <inputtype="password"name="Password2"size="20"
    tabindex="3">
    </b>
  </td>
</tr>
<tr>
    <td width="23%">
    <b>
    First Name*:</b></td>
<td width="77%">
    <b> <input type="text"name="FirstName"size="20"tabindex="4">
    </b>
</td>
</tr>
<tr>
  <td width="23%">
<b>
    Middle Name:</b> </td>
<td width="77%">
    <b>
    <input type="text" name="MiddleName" size="20" tabindex ="5">
    </b>
</td>
</tr>
<tr>
<td width="23%">
    <b>
    Last Name*:</b> </td>
<td width="77%">
    <b> <input type="text"name="LastName"size="20" tabindex="6">
```

```
    </b>
</td>
</tr>
</table>
</div>
<div align="center">
<center>
    <table border="0" width="100%"height="201">
<tr>
        <td width="23%" height="25"> <b>Address1*:  </b> </td>
        <td width="77%" height="25"> <b> <input type="text" name="Address1"
        size="43" tabindex="7"> </b> </td>
</tr>
<tr>
        <td width="23%" height="25"> <b>Address2: </b> </td>
        <td width="77%" height="25"> <b> <input type="text" name="Address2"
        size="43" tabindex="8"> </b> </td>
</tr>
<tr>
        <td width="23%" height="25"> <b>City*: </b> </td>
        <td width="77%" height="25"> <b> <input type="text" name="City"
        size="20" tabindex="9"> </b> </td>
</tr>
<tr>
        <td width="23%" height="25"> <b>State*:</b> </td>
        <td width="77%" height="25"> <b> <select size="1" name="State"
        tabindex="10">
        <option selected value="Select State">Select State</option>
        <option>Assam</option>
        <option>Arunachal Pradesh</option>
        <option>Haryana</option>
        <option>Maharashtra</option>
```

```
        <option>Rajasthan</option>
        <option>Uttar pradesh</option>
        </select> </b> </td>
    </tr>
    <tr>
        <td width="23%" height="25"> <b>Pin Code*: </b> </td>
        <td width="77%" height="25"> <b> <input type="text" name="PinCode"
        size="20" tabindex="11"> </b> </td>
</tr>
<tr>
        <td width="23%" height="25"> <b>Email ID*:</b> </td>
        <td width="77%" height="25"> <b> <input type="text" name="Email"
        size="20" tabindex="12"> </b> </td>
</tr>
<tr>
        <td width="23%" height="25"> <b>Phone Number*:</b> </td>
        <td width="77%" height="25"> <b> <input type="text" name="Phone" size
        ="20" tabindex="13"> </b> </td>
 </tr>
 <tr>
        <td width="77%" height="1"> </td>
</tr>
<tr>
        <td width="23%" height="21"> </td>
        <td width="77%" height="21">
<b>
        <input type="submit"value="Submit"name="B1"onclick="validate();"
        tabindex="13">
</b>
</td>
        </tr>
</table>
```

```
</center>
</div>
</form>
</body>
</HTML>
```

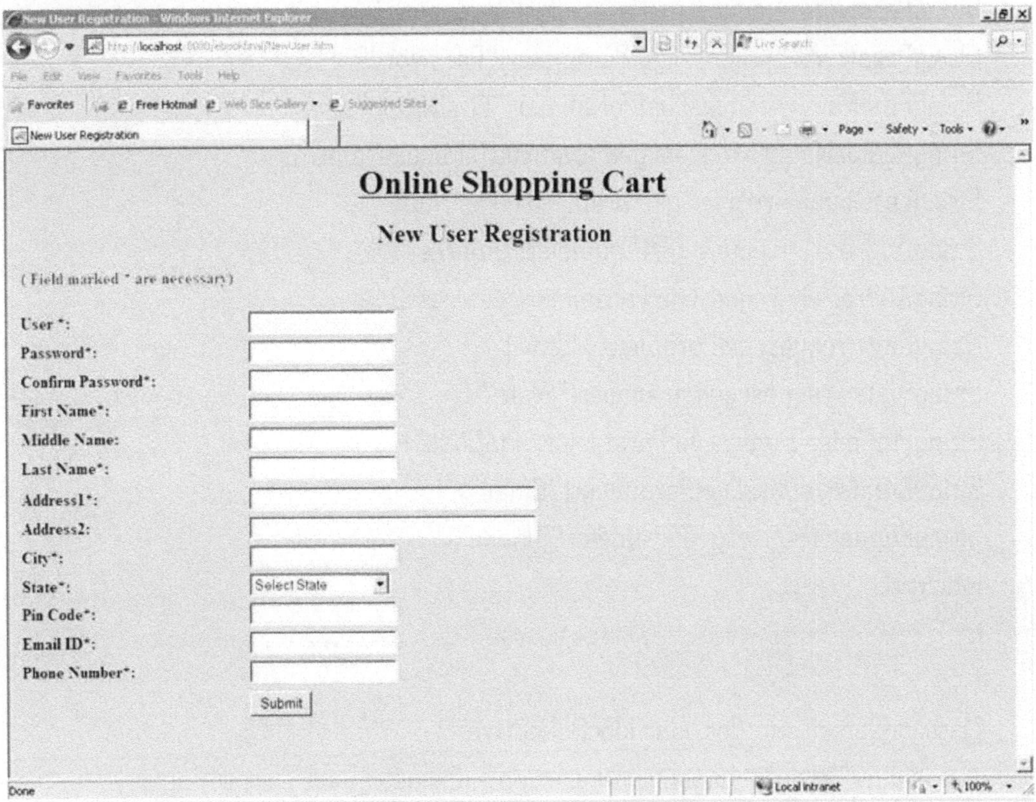

Figure 15.3 Output of New user.htm

This screen (Figure 15.3) show login form. Now fill all the entries and press submit.

Sign.jsp

```
<html>
<bodybgcolor="#CCFFFF">
<%@pageerrorPage="errorpage.jsp"language="java"import="java.sql.*"%>
< %
```

```
        Connectionconn;
        conn=null;
        ResultSet rs;
        rs=null;
    StringUserName=request.getParameter("UserName");
    StringPassword=request.getParameter("Password");
    StringPassword2=request.getParameter("Password2");
    StringFirstName=request.getParameter("FirstName");
    StringMiddleName=request.getParameter("MiddleName");
    StringLastName=request.getParameter("LastName");
    StringAddress1=request.getParameter("Address1");
    StringAddress2=request.getParameter("Address2");
    StringCity=request.getParameter("City");
    StringState=request.getParameter("State");
    StringPinCode=request.getParameter("PinCode");
    StringEmail=request.getParameter("Email");
    StringPhone=request.getParameter("Phone");
    intflag=0;
    try
    {
    Class.forName("sun.jdbc.odbc.JdbcOdbcDriver");
        conn=DriverManager.getConnection("jdbc:odbc:Data","scott","tiger");
    PreparedStatement stat= conn.prepareStatement("SELECT * FROM user_profile
        where username=?" );
        stat.setString(1,UserName);
        rs = stat.executeQuery();
    out.println("query executed");
        if(rs.next())
        {
        flag=1;
        %>
```

```
        <SCRIPT language="JavaScript">
        alert("User name already exists");
        location.href="NewUser.htm";
        </SCRIPT>
        < %
        }
      else
      {
        flag=0; //conn.close();
        }
    }
    catch(Exception E)
    {
        out.println("Error"+E);
    }
    if(flag = = 0)
    {
        try
        {
    PreparedStatement stat1=conn.prepareStatement("INSERT INTO user_auth
        values(?,?)");
        stat1.setString(1,UserName);
        stat1.setString(2,Password);
        PreparedStatement stat=conn.prepareStatement("INSERT INTO user_profile
        values(?,?,?,?,?,?,?,?,?,?,?,?)");
        stat.setString(1,UserName);
        stat.setString(2,Password);
        stat.setString(3,FirstName);
        stat.setString(4,MiddleName);
        stat.setString(5,LastName);
        stat.setString(6,Address1);
```

```
        stat.setString(7,Address2);
        stat.setString(8,City);
        stat.setString(9,State);
        stat.setString(10,PinCode);
        stat.setString(11,Email);
        stat.setString(12,Phone);
out.println("here2");
        stat.executeUpdate();
stat1.executeUpdate();
        flag=0;
        response.sendRedirect("Login.htm");
}
catch(Exception E)
{
        out.println("Error inserting value"+E);
}
finally
{
        rs.close();
        conn.close();
}
% >
</body>
</html>
```

SearchCriteria.jsp

```
<html>
<head>
<title>Search Items</title>
</head>
<body bgcolor="#CCFFFF">
<center>
```

```
<H1> <u>Online Shopping Cart</u> </H1>
</center>
<script language="JavaScript">
function startSearch()
{
      if(document.frm.R1[1].checked)
      {
            if(document.frm.BookTitle.value= ="")
            {
            alert("You must enter a book title");
            return;
      }
}
if(document.frm.R1[2].checked)
{
      if(document.frm.BookAuthor.value= ="")
      {
      alert("You must enter an author name");
      return;
      }
}
      if(document.frm.R1[3].checked)
{
      if(document.frm.BookPublisher.value= ="")
{
      alert("You must enter a publisher name");
      return;
      }
}
document.frm.action="search.jsp";
document.frm.submit();
```

```
}
</script>
< %@pageerrorPage="errorpage.jsp"import="java.net.*" % >
< %@pageimport="java.io.*" % >
< %@pageimport="java.sql.*" % >
< %
Connectioncon=null;
PreparedStatementstat=null;
ResultSetrs=null;
intctr=0,flag=0;
try
{
    Class.forName("sun.jdbc.odbc.JdbcOdbcDriver");
    con = DriverManager.getConnection("jdbc:odbc:Data","scott","tiger");
    stat = con.prepareStatement("select * from category_details");
rs=stat.executeQuery();
    %>
<form name ="frm" method="POST">
<center> <font size="5">Search Books</font> </p> </center>
<BR> <BR>
<table border="0" width="100%" height="196">
<tr>
<td width="17%" align="right" height="25"> <input type="radio"
    value="Category" checked name="R1"> </td>
    <td width="32%" height="25">Search by Category</td>
    <td width="51%" height="25">
        <select size="1" name="BookCategory">
        <option selected value="Select">Select Category</option>
            <%
            while(rs.next())
            {
```

```
                    String category=rs.getString(2);%>
                    <option value="< %=category%>"> <%=category%> </option>
                    <%}%>
        </select>
    </td>
    </tr>
    <tr>
        <td width="17%" align="right" height="25"> <input type="radio"
        name="R1" value="Title"> </td>
            <td width="32%" height="25">Search by Book Title</td>
            <td width="51%" height="25"> <input type="text" name="BookTitle"
        size ="25"> </td>
    </tr>
    <tr>
        <td width="17%" align="right" height="25"> <input type="radio"
        name="R1" value="Author"> </td>
        <td width="32%" height="25">Search by Book Author</td>
        <td width="51%" height="25"> <input type="text" name="BookAuthor"
        size="25"> </td>
    </tr>
    <tr>
        <td width="17%" align="right" height="25"> <input type="radio"
        name="R1" value="Publisher"> </td>
        <td width="32%" height="25">Search by Book Publisher</td>
        <td width="51%" height="25"> <input type="text" name="BookPublisher"
        size="25"> </td>
    </tr>
    <tr>
    </tr>
    <tr>
        <td width="17%" align="right" height="21"> </td>
        <td width="32%" height="21">  
```

```
        <input type="button" value="   Search   "
    name="Search"
onClick="startSearch();"> </td>
<td width="51%" height="21"> <a href="Logout.jsp">Click here to logout</a> </
    td>
</tr>
</table>
</form>
<%
}
catch(Exception e)
{
    out.print("Error = " + e + "<HR>");
}
finally
{
    rs.close();
    con.close();
}
% >
</body>
</html>
```

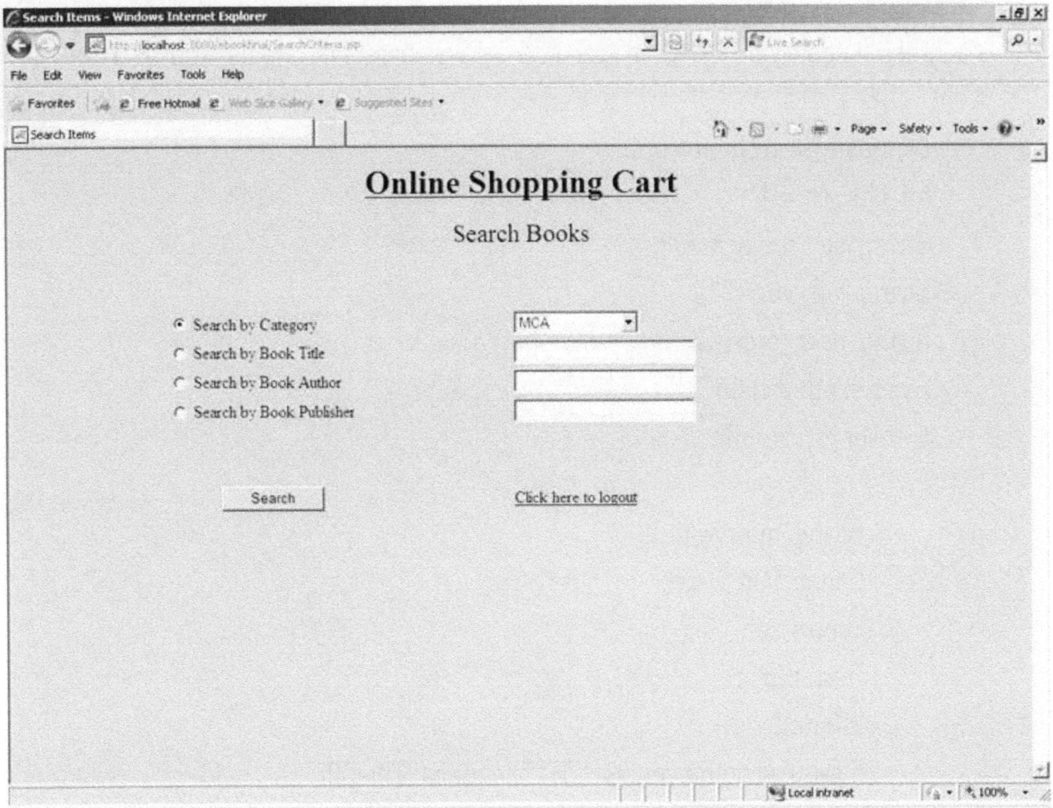

Figure 15.4 Output of Search Crteria.jsp

This screen (Figure 15.4) show all searching filter of this project.

Search.jsp

```
<%@ page errorPage="errorpage.jsp" language="java" import="java.sql.*" %>
<html>
<head>
<title>Search Results</title>
</head>
<body bgcolor="#CCFFFF">
<center>
<H1> <u>Online Shopping Cart</u> </H1>
<font size="5" >Search Results</font>
</center>
```

```
<BR>
<a href="Logout.jsp" > <font size="4">Click here to Logout</font></a>
< %
    Stringstr_searchvar=null;
    int chk_ctr=0;
    String str_colvar=null;
    String tab_var="";
    String user_src=(String)session.getValue("user");
    System.out.println("0");
    if (user_src!=null)
    {
        String query=null;
        Connection conn;
        conn=null;
        ResultSet rs;
        rs=null;
        if(request.getParameter("R1").equals("Title"))
        {
            str_searchvar=request.getParameter("BookTitle");
            str_colvar="TITLE";
        }
        if(request.getParameter("R1").equals("Category"))
        {
            chk_ctr=1;
            str_searchvar=request.getParameter("BookCategory");
            str_colvar="Categoryname";
}
    if(request.getParameter("R1").equals("Publisher"))
{
    str_searchvar=request.getParameter("BookPublisher");
    str_colvar="Publisher";
```

```
}
if(request.getParameter("R1").equals("Author"))
{
        str_searchvar=request.getParameter("BookAuthor");
        str_colvar="Author";
}
if (request.getParameter("R2")==null)
{
        tab_var="BOOK_DETAILS";
}
else
{
// to get the name of the table in which the search to be done-------
        if (request.getParameter("R2").equals("A"))
        {
                tab_var="TEMP_DETAIL";
}
else
{
                tab_var="BOOK_DETAILS";
        }
}
System.out.println(str_searchvar);
query = "select a.* from "+tab_var+" a, category_details b where a."+str_colvar+"
        = ? and a.Categoryid = b.Categoryid";
System.out.println("1");
try
{
Class.forName("sun.jdbc.odbc.JdbcOdbcDriver");
conn =DriverManager.getConnection("jdbc:odbc:Data","scott","tiger");
int ctr=0;
```

```
System.out.println("2");
PreparedStatement stm = conn.prepareStatement(query);
stm.setString(1,str_searchvar);
rs = stm.executeQuery();
%>
     <form name="cart" action="Inter_Cart.jsp" method="POST">
     <table border=1>
     < tr>
     <td>S.No</td>
     <td>Select</td>
     <td>Book Name</td>
     <td>Author Name</td>
     <td>Publisher</td>
     <td>Edition</td>
     <td>Price (in Rs.)</td>
     <td>Quantity Available(Nos.)</td>
     <td>Description</td>
     </tr>
< %
     while(rs.next())
     {
     System.out.println("inside while");
     ctr=ctr+1;
out.println("<tr> <td>"+ctr+"</td>");
int int_Bookid=0;
int_Bookid=rs.getInt(1);
System.out.println(int_Bookid);
out.println("<td> <input type=checkbox name=chk"+ctr+"
     value="+int_Bookid+">");
int int_Categoryid=rs.getInt(2);
String str_TITLE=rs.getString(3);
```

```
out.println("<td>");
out.println(str_TITLE);
out.println("</td>");
String str_Author=rs.getString(4);
out.println("<td>");
out.println(str_Author);
out.println("</td>");
String str_Publisher=rs.getString(5);
out.println("<td>");
out.println(str_Publisher);
out.println("</td>");
String str_Edition=rs.getString(6);
out.println("<td>");
out.println(str_Edition);
out.println("</td>");
int int_PRICE=rs.getInt(7);
out.println("<td>");
out.println(int_Price);
out.println("</td>");
int int_Quantity=rs.getInt(8);
out.println("<td>");
out.println(int_Quantity);
out.println("</td>");
String str_Description=rs.getString(9);
out.println("<td>");
out.println(str_Description);
out.println("</td> </tr>");
String ins_query ="Insert Into temp_detail Values (?,?,?,?,?,?,?,?,?)";
PreparedStatement stat_ins= conn.prepareStatement(ins_query);
stat_ins.setInt(1,int_Bookid);
stat_ins.setInt(2,int_Categoryid);
```

```
stat_ins.setString(3,str_Title);
stat_ins.setString(4,str_Author);
stat_ins.setString(5,str_Publisher);
stat_ins.setString(6,str_Edition);
stat_ins.setInt(7,int_Price);
stat_ins.setInt(8,int_Quantity);
stat_ins.setString(9,str_Description);
int x = stat_ins.executeUpdate();
System.out.println(x);
}
    String str_ctr=String.valueOf(ctr);
    session.putValue("ctr_val",str_ctr);
%>
    </table>
    <br>
    <input type="submit"value="Add to Cart"name="Add">
    </form>
<%
rs.close();
stm.close();
}
catch(Exception e){}
}
else
    {
        response.sendRedirect("Unauthorised.htm");
    }
%>
</body>
</html>
```

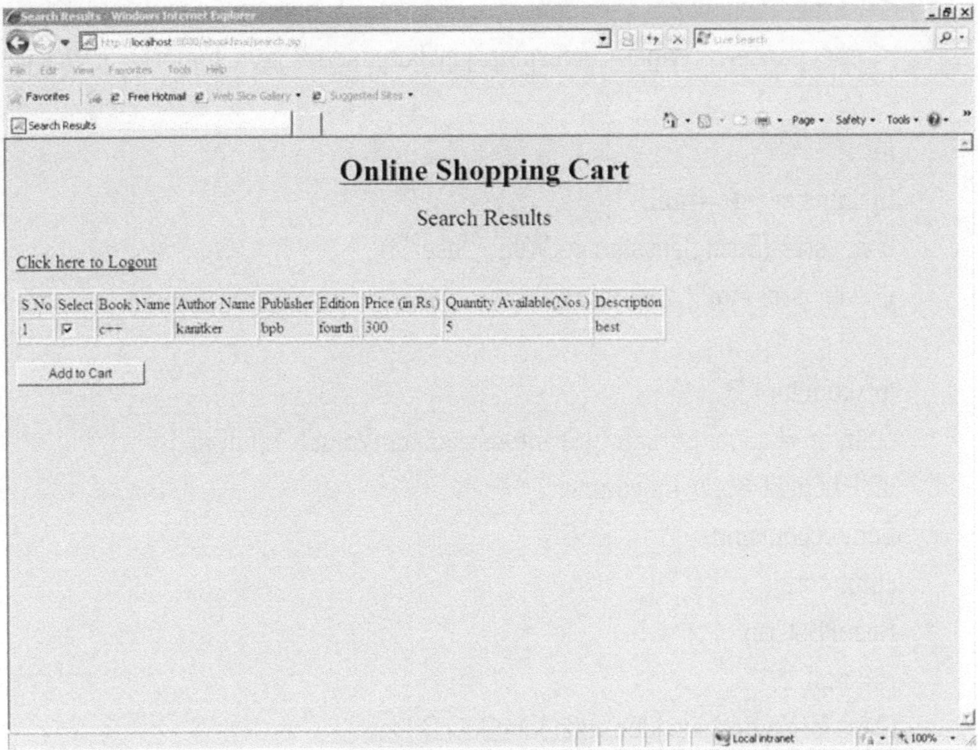

Figure 15.5 Output of Search.jsp

This screen (Figure 15.5) display search results of the project.

Cart.jsp

```
<html>
<head>
<title>ShoppingCart</title>
</head>
<body bgcolor="#CCFFFF">
    <center>
    <H1> <u>OnlineShoppingCart</u> </H1>
    </center>
    <BR>
```

```
<a href="Logout.jsp" > <font size="4">Click here to Logout</font> </a>
< %@pageerrorPage="errorpage.jsp"language="java"import="java.sql.*"%>
< %
int i;
Stringuser_src=null;
user_src=(String)session.getValue("user");
if(user_src!=null)
{
int counter=0;
counter=Integer.parseInt((String)session.getValue("TotalSel"));
int[] bookid=new int[counter];
Connectionconn;
conn=null;
ResultSet rs;
rs=null;
Class.forName("sun.jdbc.odbc.JdbcOdbcDriver");
conn =DriverManager.getConnection("jdbc:odbc:Data","scott","tiger");
PreparedStatement stat=null;
int ctr=0,bookprice=0;
for(i=1;i<=counter;i++)
{
    if((String)session.getValue("chk_var"+i)!=null)
    {
bookid[i-1]=Integer.parseInt((String)session.getValue("chk_var"+i));
}
}
session.putValue("BookArr",bookid);
%>
<h3>YourCartDetailsareasfollows:</h3>
<formname="shop"action="Cart_Quantity.jsp">
<tableborder=1>
```

```
    <tr>
    <td>S.No</td>
    <td>BookName</td>
    <td>Price(in Rs.)</td>
    </tr>
    < %
    int price=0;
    for (int j=1;j <=counter; j++)
    {
          out.println("<tr><td>"+j);
          String query="select title,pricefrombook_detailswherebookid=?";
          stat= conn.prepareStatement(query);
          stat.setInt(1,bookid[j-1]);
rs = stat.executeQuery();
while (rs.next())
{
    out.println("<td>");
          out.println(rs.getString(1));
          out.println("<td>");
          price=rs.getInt(2);
          out.println(price);
    }
    rs.close();
    bookprice=bookprice+price;
    session.putValue("bPrice",String.valueOf(bookprice));
    out.println("</tr>");
}
ResultSet rscombo=null;
String query1="select * from category_details";
PreparedStatement stat_combo = conn.prepareStatement(query1);
rscombo=stat_combo.executeQuery();%>
```

```
<tr>
    <td colspan=3>Total Cart Amount(in Rs.)=
    <%=bookprice%> <td>
</tr>
</table>
<br>
<input type=submit value="Place Order">
</form>
<br> <hr> <h3>Search More</h3>
<form name="search" action="NewSearch.jsp" method="POST">
<table border=1>
<tr>
    <td> <input type="radio" value="Category" checked name="R1"> </td>
    <td>Search based on Category</td>
    <td>
    <select size="1" name="BookCategory">
    <option selected value="Select">Select Category</option>
        <%
        while(rscombo.next())
        {
         String category=rscombo.getString(2);
    %>
        <option value="<%=category%>"><%=category%></option>
        <%}%>
    </select>
    </td>
</tr>
<tr>
    <td> <input type="radio" name="R1" value="Title"> </td>
    <td>Search by Book Title</td>
    <td> <input type="text" name="BookTitle" size="25"> </td>
```

```
</tr>
<tr>
<td> <input type="radio" name="R1" value="Author"> </td>
    <td>Search by Book Author</td>
    <td> <input type="text" name="BookAuthor" size="25"> </td>
</tr>
<tr>
<td> <input type="radio" name="R1" value="Publisher"> </td>
    <td>Search by Book Publisher</td>
    <td> <input type="text" name="BookPublisher" size="25"> </td>
</tr>
<tr>
    <td> <input type="radio" name="R2" value="A">Advanced Search</td>
        <td> <input type="radio" name="R2" value="N" checked>New
    Search</td>
        <td> <input type=submit value="Search"> < /td>
    </tr>
    </table>
    </form>
    < %}
else
{
    response.sendRedirect("Unauthorised.htm");
}
    % >
</body>
</html>
```

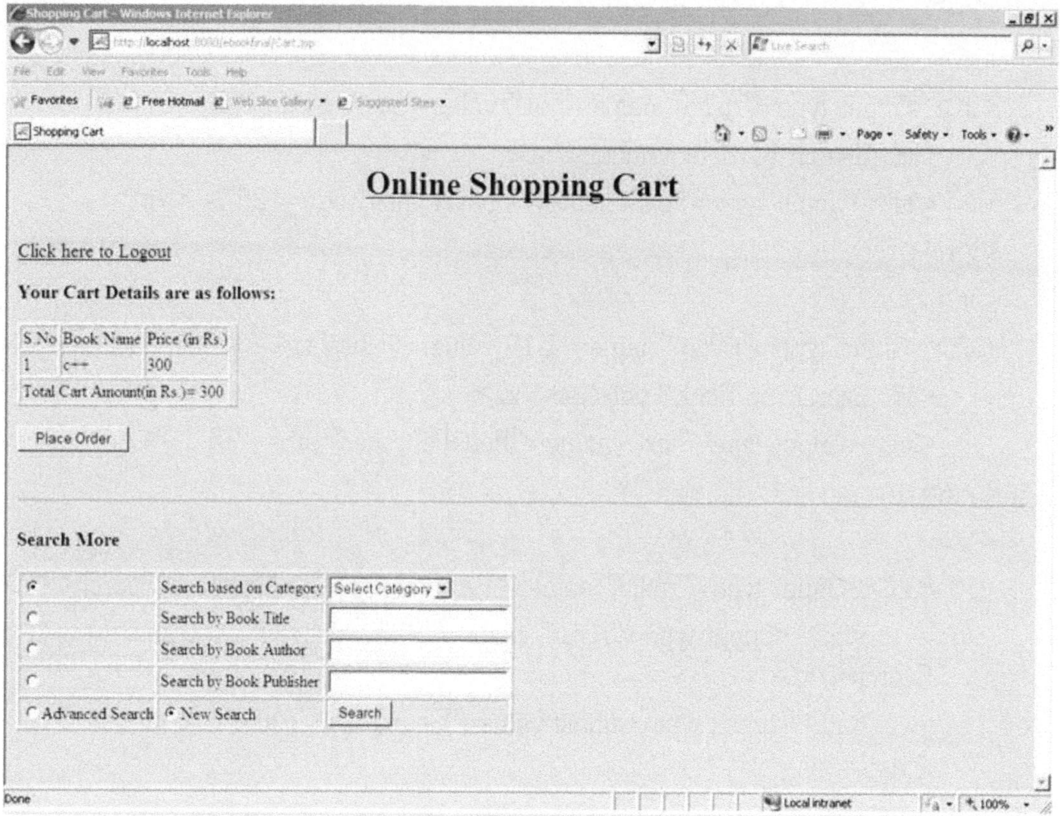

Figure 15.6 Output of Cart.jsp

This screen (Figure 15.6) display previous search result and new search filter.

Cart_Quantity.jsp

```
<html>
<head>
<title>ShoppingCart</title>
</head>
<body bgcolor="#CCFFFF">
    <center>
    <H1> <u>OnlineShoppingCart</u> </H1>
    </center>
```

```
    <BR>
<a href="Logout.jsp"> <font size="4">ClickheretoLogout</font> </a>
<%@pageerrorPage="errorpage.jsp"language="java"import="java.sql.*"% >
< %
int i;
Stringuser_src=null;
user_src=(String)session.getValue("user");
if(user_src!=null)
{
    int counter=0;
    counter=Integer.parseInt((String)session.getValue("TotalSel"));%> <script
    language="Javascript">
function check_sum()
{
    var cost=0;
    x_avlqty=new Array(<%=counter%>);
    y_orderqty=new Array(<%=counter%>);
    z_price=new Array(<%=counter%>);
    for (var i=0;i<<%=counter%>;i++)
{
        z_price[i](document.shop.elements[i*4].value);
        x_avlqty[i]=document.shop.elements[i*4+1].value;
        y_orderqty[i]=document.shop.elements[i*4+2].value;
        if(parseInt(y_orderqty[i])<0)
{
    alert("Book quantity cannot be Negative");
    return false;
}
if (parseInt(x_avlqty[i])<parseInt(y_orderqty[i]))
        {
            alert("Order Quantity Cannot be greater than available quantity");
```

```
                    return false;
                }
                else
                {
document.shop.elements[i*4+3].value=((parseInt(y_orderqty[i]))*(parseInt(z_price[i])));
cost=cost+parseInt(document.shop.elements[i*4+3].value);
                }
            }
document.shop.tot_val.value=cost;
}
</script>
<%
        int[] bookid=newint[counter];
        Connectionconn;
        conn=null;
        ResultSet rs;
        rs=null;
        Class.forName("sun.jdbc.odbc.JdbcOdbcDriver");
        conn =DriverManager.getConnection("jdbc:odbc:Data","scott","tiger");
PreparedStatement stat=null;
int ctr=0,bookprice=0;
for(i=1;i<=counter;i++)
{
        if((String)session.getValue("chk_var"+i)!=null)
        {
bookid[i-1]=Integer.parseInt((String)session.getValue("chk_var"+i));
        }
        }
        session.putValue("BookArr",bookid);
        %>
            <h3>YourCartDetailsareasfollows:</h3>
```

```
<form name="shop" action="Shop.jsp" method="POST">
<table border=1>
<tr>
<td>S.No</td>
<td>Book Name</td>
<td>Price per unit (in Rs.)</td>
<td>Quantity Available</td>
<td>Quantity Purchased</td>
<td>Amount</td>
</tr>
< %
int price=0;
int qty_avbl=0;
for (int j=1;j<=counter;j++)
{
    out.println("<tr><td>"+j);
    out.println("</td>");
String query="select title, price,quantity from book_details where bookid=?";
stat= conn.prepareStatement(query);
stat.setInt(1,bookid[j-1]);
rs = stat.executeQuery();
while (rs.next())
{
    out.println("<td>");
    out.println(rs.getString(1));out.println("</td>");
    price=rs.getInt(2);%>
<td> <input type="hidden"name="bk_price<%=j%>"value=<%=price%>
</td>    <%out.println(price);
qty_avbl=rs.getInt(3);%>
<td> <input type=hidden name="qty_avl<%=j%>"
value=<%=qty_avbl%>> <%=qty_avbl%> </td>
```

```
<td> <input type=text name="qty<%=j%>" value="0" onChange="return
check_sum();"> </td>
      <td> <input type=text name="cal_amt<%=j%>"> </td>
<%
}%>
</tr>
<%
rs.close();
bookprice=bookprice+price;
session.putValue("bPrice",String.valueOf(bookprice));
      out.println("</tr>");
}%>
<tr> <td> <td> <td> <td> <td>Total Amount(in Rs.)<td> <input
type="text" name="tot_val" value="0"> </td> </tr>
</table>
<br>
<input type=submit value="Place Order">
</form>
<br> <hr>
<%}
else
      {
            response.sendRedirect("Unauthorised.htm");
      }
% >
      </body>
</html>
```

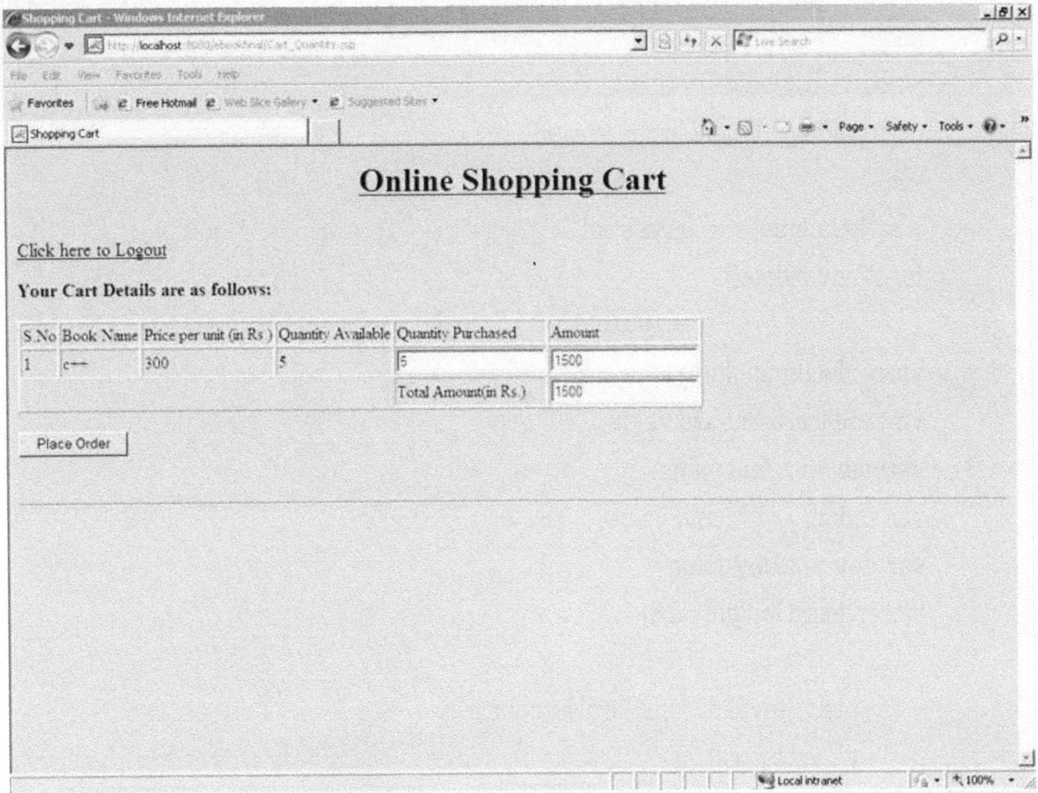

Figure 15.7 Output of Cart Quantity.jsp

This screen (Figure 15.7) display bill information of the purchased book

Shop.jsp

```
<html>
<head>
<title>OrderDetails</title>
</head>
<bodybgcolor="#CCFFFF">
    <center>
    <H1> <u>OnlineShoppingCart</u> </H1>
    </center>
    <BR>
    <a href="Logout.jsp"> <font size="4">Click here to Logout</font> </a>
```

```
<%@ page errorPage="errorpage.jsp" language="java" import="java.sql.*"
%>
< %
if (session.getValue("user")!=null)
{% >
<scriptlanguage="Javascript">
functionvalidate()
{
varr=document.form1;
varcreditcard=r.Card.value;
varyear = r.Year.value;
var month = r.Month.value;
var day = r.Day.value;
if(creditcard.length!=16)
{
        alert("InvalidCreditCardNumber");
        r.Card.focus();
        return;
    }
//-------------------------------------------------------------------
if(day=="Select Day")
{
    alert("Please select the day");
    r.Day.focus();
    return;
}
//-------------------------------------------------------------------
if(month=="Select Month")
{
        alert("Please select the month");
        r.Month.focus();
```

```
            return;
        }
    //---------------------------------------------------------------
        if(year.length==0)
        {
            alert("Please enter the year");
            r.Year.focus();
    }
    elseif(year.length!=4)
    {
        alert("Please enter the year in YYYY format");
        r.Year.focus();
    }
    //---------------------------------------------------------------
        if (( month = = "3"||month= = "5"||month= ="8"||month= ="10")&&(day=
        ="31"))
        {
    alert("Please enter a valid date");
        r.Day.focus();
    }
    if ((month = = "1") && (day = = "30" || day = = "31"))
    {
        alert("Please enter a valid date");
        r.Day.focus();
    }
        //------- Tocheckthatcarddateisnotlessthancurrentdate----------
    varvr_day;
    varvr_month;
    varvr_year;
    vard=newDate();
    vr_year=d.getFullYear();
```

```
vr_month=d.getMonth();
vr_day=d.getDate();
if(parseInt(vr_year)>parseInt(document.form1.Year.value))
{
	alert("YearcannotbelessthanCurrentYear");
	returnfalse;
}
elseif(parseInt(vr_year)==parseInt(document.form1.Year.value))
{
	if(parseInt(vr_month)>parseInt(document.form1.Month.value))
	{
		alert("MonthcannotbelessthanCurrentMonth");
		return false;
}
else if(parseInt(vr_month)==parseInt(document.form1.Month.value))
{
	if(parseInt(vr_day)>parseInt(document.form1.Day.value))
	{
		alert("Day cannot be less than current Day");
		return false;
		}
	}
	}
//-----------------------------------------------------------------
	document.form1.submit();
document.forms[0].action="Final.jsp";
}
</script>
<%
	int counter=0;
	counter=Integer.parseInt((String)session.getValue("TotalSel"));
```

```
        String query="";
        int[] bookid=new int[counter];
        int[] bk_price=new int[counter];
        int[] bk_quantity=new int[counter];
    Connection conn;
        conn=null;
        ResultSet rs;
        rs=null;
        Class.forName("sun.jdbc.odbc.JdbcOdbcDriver");
        conn =DriverManager.getConnection("jdbc:odbc:Data","scott","tiger");
        PreparedStatement stat=null;
        int ctr=0,bookprice=0;
        int i=1;
        while(i<=counter)
        {
                if((String)session.getValue("chk_var"+i)!=null)
                {
    bookid[i-1]=Integer.parseInt((String)session.getValue("chk_var"+i));
        query="Select Price From Book_Details Where Bookid=?";
                stat=conn.prepareStatement(query);
        stat.setInt(1,bookid[i-1]);
        rs=stat.executeQuery();
        if(rs.next())
        {
                bk_price[i-1]=rs.getInt(1);
        }
        else
        {
                bk_price[i-1]=0;
        }
    }
```

```
if(request.getParameter("qty"+i)!=null)
{
      bk_quantity[i-1]=Integer.parseInt(request.getParameter("qty"+i));
            String sess_var=String.valueOf(bk_quantity[i-1]);
            session.putValue(sess_var,String.valueOf(bk_quantity[i-1]));
      }
bookprice=bookprice+(bk_price[i-1]*bk_quantity[i-1]);
            i++;
}
      session.putValue("bPrice",String.valueOf(bookprice));
%>
<form method="POST" action="Final.jsp" name="form1">
<table width="70%">
<tr>
      <td colspan=4> <b>TotalAmount(inRs.)is :<%=bookprice%> </b> <td> </
      tr>
<tr>
</tr>
<tr>
      <td> <font size="3"> <b>CreditCardNumber:</b> </font> </td>
      <td colspan=3>
<inputtype="text"name="Card"size="39"maxlength="16"tab="1"> </td>
</tr>
<tr>
      <td> <font size="3"> <b> ExpiryDate:</b> </font> </td>
      <td>
<select size="1" name="Day">
<option value="Select Day">Select Day</option>
<option value="1">1</option>
<option value="2">2</option>
<option value="3">3</option>
```

```
<option value="4">4</option>
<option value="5">5</option>
<option value="6">6</option>
<option value="7">7</option>
<option value="8">8</option>
<option value="9">9</option>
<option value="10">10</option>
<option value="11">11</option>
<option value="12">12</option>
<option value="13">13</option>
<option value="14">14</option>
<option value="15">15</option>
<option value="16">16</option>
<option value="17">17</option>
<option value="18">18</option>
<option value="19">19</option>
<option value="20">20</option>
<option value="21">21</option>
<option value="22">22</option>
<option value="23">23</option>
<option value="24">24</option>
<option value="25">25</option>
<option value="26">26</option>
<option value="27">27</option>
<option value="28">28</option>
<option value="29">29</option>
<option value="30">30</option>
<option value="31">31</option>
</select>
</td>
<td>
```

```
<select size="1" name="Month">
<option value="Select Month">Select Month</option>
<option value="0">January</option>
<option value="1">February</option>
<option value="2">March</option>
<option value="3">April</option>
<option value="4">May</option>
<option value="5">June</option>
<option value="6">July</option>
<option value="7">August</option>
<option value="8">September</option>
<option value="9">October</option>
<option value="10">November</option>
<option value="11">December</option>
</select>
</td>
<td>
<font size="3"> <input type="text" name="Year" size="11"> </font>
</td>
</tr>
</table>
</font> <input type="button" value="Submit" name="B1" onClick="validate();" >
</form>
<%
}
else
{
    response.sendRedirect("Unauthorised.htm");
}%>
</body>
</html>
```

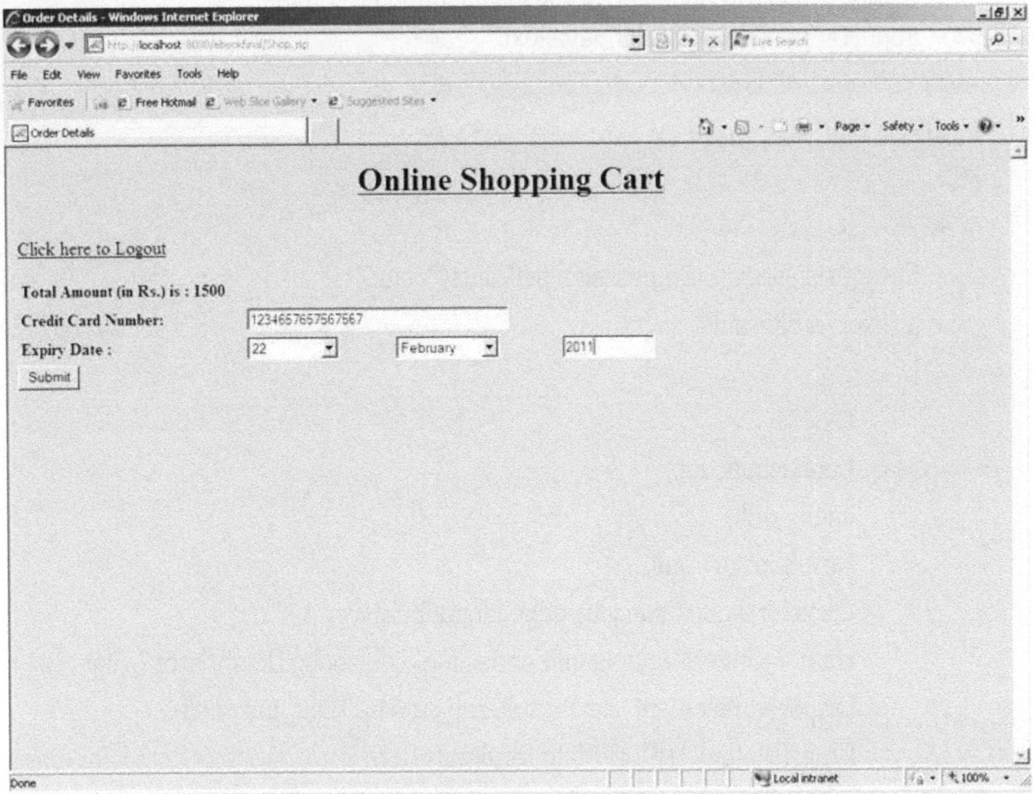

Figure 15.8 Output of shop.jsp

This screen (Figure 15.8) display credit card information of user.

Final.jsp

```
<html>
<head>
<title>Orderplaced</title>
</head>
<bodybgcolor="#CCFFFF">
    <center>
    <H1> <u>OnlineShoppingCart</u> </H1>
    <H2>OrderDetails</H2></center>
    <BR>
    <a href="Logout.jsp"> <font size="4">ClickheretoLogout</font> </a>
```

```
    <%@pageerrorPage="errorpage.jsp"language="java"
    import="java.sql.*,java.util.*,java.text.*"  %>
<%!
int OrderID,price;
%>
<%
    Stringuser_src=(String)session.getValue("user");
    if (user_src!= null)
    {
        try{
        Connectionconn;
        conn=null;
        ResultSet rs=null;
        Class.forName("sun.jdbc.odbc.JdbcOdbcDriver");
        conn =DriverManager.getConnection("jdbc:odbc:Data","scott","tiger");
        PreparedStatement stat1=null,stat_sel=null,stat_ins=null;
        OrderID=(int)(10000*Math.random()+1);
        price=Integer.parseInt((String)session.getValue("bPrice"));
        intcounter=0;
        counter=Integer.parseInt((String)session.getValue("TotalSel"));
        int[]bookid=new int[counter];
        int[]bk_quantity=new int[counter];
        intorg_bkqty=0;
        inti=0;
        Stringins_query="";
        Stringsel_qtyQuery="";
        Stringins_qty="";
        for(i=1;i<=counter;i++)
    {
        ResultSet rs_sql=null;
        if((String)session.getValue("chk_var"+i)!=null)
```

```
{
bookid[i-1]=Integer.parseInt((String)session.getValue("chk_var"+i));
}
if((String)session.getValue("bookqty"+i)!=null)
{
bk_quantity[i-1]=Integer.parseInt((String)session.getValue("bookqty"+i));
}
ins_query="Insertintoorder_Detailsvalues(?,?,?)";
stat1=conn.prepareStatement(ins_query);
stat1.setInt(1,OrderID);
stat1.setInt(2,bookid[i-1]);
stat1.setInt(3,bk_quantity[i-1]);
int rs_int=0;
rs_int=stat1.executeUpdate();
sel_qtyQuery="Selectquantityfrombook_Detailswherebookid=?";
stat_sel = conn.prepareStatement(sel_qtyQuery);
stat_sel.setInt(1,bookid[i-1]);
rs_sql=stat_sel.executeQuery();
if(rs_sql.next())
{
    org_bkqty=rs_sql.getInt(1);
}
rs_sql.close();
Stringns_qty="updatebook_detailssetquantity=?where bookid=?";
stat_ins=conn.prepareStatement(ns_qty);
stat_ins.setInt(1,org_bkqty-bk_quantity[i-1]);
stat_ins.setInt(2,bookid[i-1]);
rs_int=stat_ins.executeUpdate();
}
    }catch(Exception e){}
%>
```

```
<form name="newsrc"  method="POST">
<BR> <font size="3"> <b>
Your order has been successfully placed.
<BR> <BR>Order Number is : <%=OrderID%>
<BR> <BR>Total Amount is : <%=price%>
<BR> <BR>Date of Order is:
< %
out.println((new  java.util.Date()).toLocaleString());
%>
<BR> <BR>
        Your Order will be delivered within next 48 working hours
< %
    String  card_no=request.getParameter("Card");
    Connection con= null;
    Class.forName("sun.jdbc.odbc.JdbcOdbcDriver");
    con =DriverManager.getConnection("jdbc:odbc:Data","scott","tiger");
    java.util.Datenow=newjava.util.Date();
    DateFormatdf1=DateFormat.getDateInstance(DateFormat.SHORT);
Strings1=df1.format(now);
try
{
Stringstr= "insertintoorder_tablevalues(?,?,?,?)";
PreparedStatement stat= con.prepareStatement(str);
stat.setInt(1,OrderID);
stat.setString(2,user_src);
stat.setInt(3,price);
stat.setString(4,s1);
int x = stat.executeUpdate();
}catch(Exception e){out.println(e); }
}
else
```

```
{
        response.sendRedirect("Unauthorised.htm");
        }
        %>
</body>
</html>
```

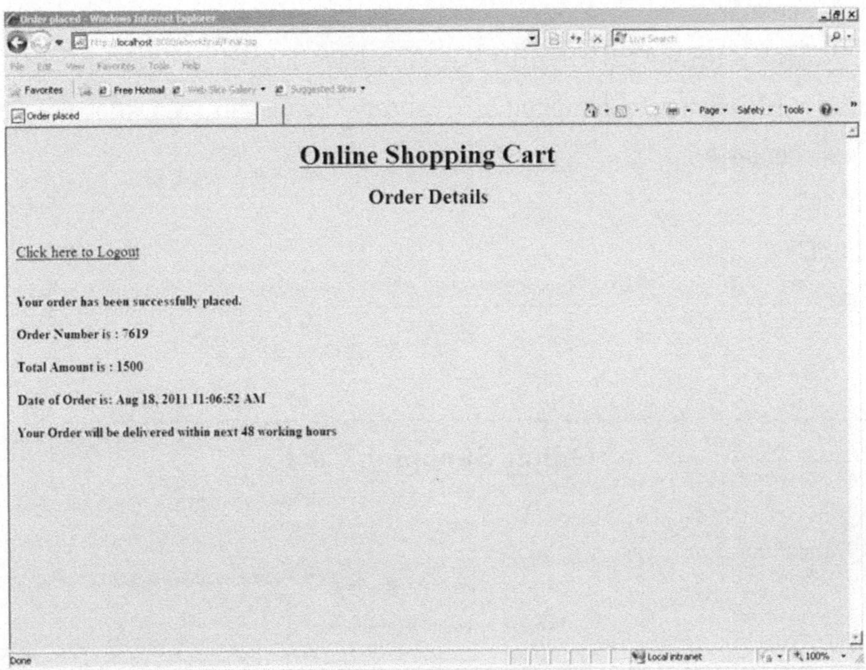

Figure 15.9 Output of Final.jsp

This screen (Figure 15.9) display total bill of the book using card.

Logout.jsp

```
<html>
<head>
<title>Logged  out</title>
</head>
<body bgcolor="#CCFFFF">
```

```
    <center>
    <H1> <u>OnlineShoppingCart</u> </H1>
    </center>
    <%@pageerrorPage="errorpage.jsp"language="java" %>
< %
    session.invalidate();
    % >
    <center>
    <BR> <BR> <BR> <BR> <b>You have logged out successfully.
<BR> <BR>ReturntoShoppingCart?<ahref="Login.htm">Clickhere</a> </b>
    </center>
</body>
</html>
```

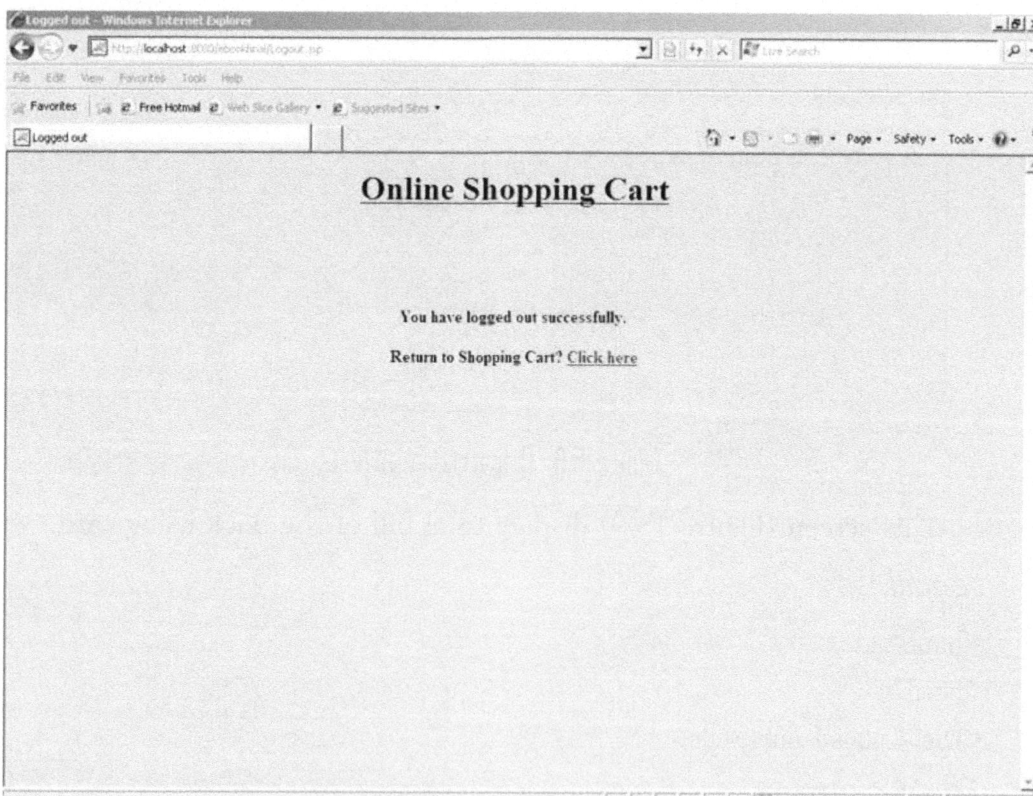

Figure 15.10 Output of Logout.jsp

This screen (Figure 15.10) display logout and relogin information of the project.

admin.htm

```
<html>
<head>
<title>administeringshoppingcart</title>
</head>
<bodybgcolor="#ccffff">
      <center>
      <h1>administeringonlineshoppingcart</h1>
      </center>
      <br><hr><br><br>
      <h2>welcomeadministrator</h2> <br>
      <formname="admin"method="post">
      <tablewidth=100%border="0"cellpadding="2"cellspacing="0">
      <tr>
            <th align = "left" width="4%"> </th>
            <th align = "left" width="20%"> </th>
            <th align = "left" width="38%"> </th>
            <th align = "left" width="38%"> </th>
</tr>
<tr>
      <th>1</th>
      <th align = "left">book</th>
      <th> <a href="insert.jsp">insert new book details</a> </th>
      <th> <a href="modify.jsp">modify book details</a> </th>
</tr>
<tr>
      <th>2</th>
      <th align = "left">category</th>
      <th> <a href="insert_category.jsp">insertbookcategory</a> </th>
```

```
<th> <a href="update_category.jsp">modifybookcategory</a> </th>
    </tr>
</body>
</html>
```

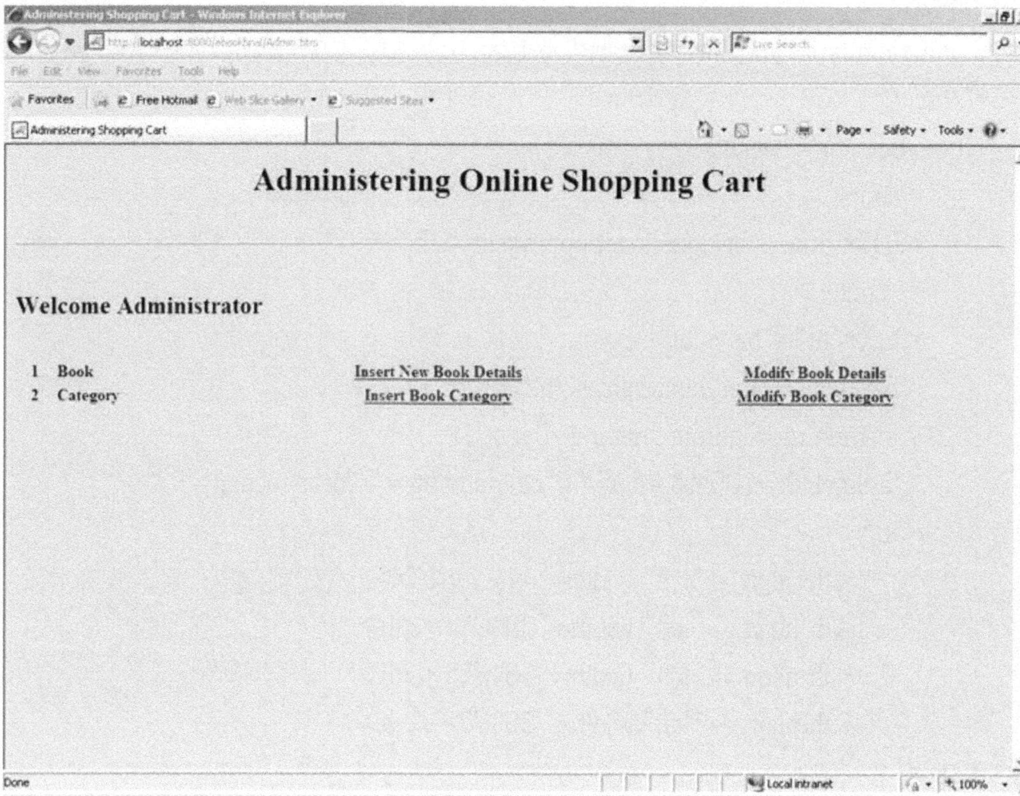

Figure 15.11 Output of Admin.htm

This screen (Figure 15.11) display book and category information of the project.

Insert.jsp

```
<html>
<head>
<title>insertnewbook</title>
</head>
<body>
```

```
<center>
<h1> <u>administeringonlineshoppingcart</u> </h1>
<h2> add book</h2>
</center>
<br>
<hr>
<br>
<a href="logout.jsp"> <font size="4">clickheretologout</font></a>
<body bgcolor="#ccffff">
<SCRIPTlanguage="Javascript">
functioncheck()
{
varctr=0;
if(document.addbook.title.value=="")
{
      alert("Please Enter Title");
      return;
}
if(document.addbook.author.value=="")
{
      alert("Please Enter Author");
      return;
}
if(document.addbook.pub.value=="")
{
      alert("PleaseEnterPublisherName");
      return;
}
if(document.addbook.edition.value=="")
{
alert("PleaseEnterEdition");
```

```
return;
}
if(document.addbook.price.value=="")
{
    alert("PleaseEnterPrice(in Rs.)");
    return;
}
if(document.addbook.qty.value=="")
{
    alert("Please Enter Quantity");
    return;
}
if(document.addbook.desc.value=="")
{
    alert("PleaseEnterDescription");
    return;
}
if(isNaN(document.addbook.price.value))
{
    alert("PleaseEnteranumericvalueforPrice(inRs.)");
    ctr=1;
}
if(isNaN(document.addbook.qty.value))
{
    alert("PleaseEnteranumericvalueforQuantity");
    ctr=1;
}
if(ctr==0)
{
    document.forms[0].action="AddBook.jsp";
    document.forms[0].submit();
```

```
}
else
{
    return true;
};
}
</SCRIPT>
<%@pageerrorPage="errorpage.jsp"import="java.net.*" %>
<%@pageimport="java.io.*" %>
<%@pageimport="java.sql.*" %>
<%
Connectioncon=null;
//Statementstat=null;
PreparedStatementstat=null;
//Statementstat1=null;
PreparedStatementstat1=null;
ResultSetrs=null;
ResultSetrs1=null;
Stringquery=null;
Stringquery1=null;
int ctr=0,flag=0;
try
{
Class.forName("sun.jdbc.odbc.JdbcOdbcDriver");
con = DriverManager.getConnection("jdbc:odbc:Data","scott","tiger");
query="Select a.*,b.categorynamefrombook_detailsa,category_detailsbwhere
    a.categoryid=b.categoryid";
query1="select * fromcategory_details";
stat = con.preparestatement(query);
stat1 = con.preparestatement(query1);
out.println("here1");
```

```
rs=stat.executequery();
rs1=stat1.executequery();
%>
<br><h3>list of available books:</h3>
<table border=1>
<tr>
      <td>s.no</td>
      <td>book id</td>
      <td>book name</td>
      <td>author name</td>
      <td>publisher</td>
      <td>edition</td>
      <td>price (in rs.)</td>
      <td>quantity available(nos.)</td>
      <td>description</td>
      <td>category</td>
      </tr>
<%
intint_bookid,int_categoryid,int_price,int_quantity;
stringstr_title,str_author,str_publisher,str_edition,str_description;
while(rs.next())
{
      ctr=ctr+1;
      out.println("<tr><td>"+ctr);
      int_bookid=rs.getint(1);
      out.println("<td>");
      out.println(int_bookid);
      int_categoryid=rs.getint(2);
      str_title=rs.getstring(3);
      out.println("<td>");
      out.println(str_title);
```

```
str_author=rs.getstring(4);
out.println("<td>");
out.println(str_author);
str_publisher=rs.getstring(5);
out.println("<td>");
out.println(str_publisher);
str_edition=rs.getstring(6);
out.println("<td>");
out.println(str_edition);
int_price=rs.getint(7);
out.println("<td>");
out.println(int_price);
int_quantity=rs.getint(8);
out.println("<td>");
out.println(int_quantity);
str_description=rs.getstring(9);
out.println("<td>");
out.println(str_description);
out.println("<td>");
out.println(rs.getstring(10));
out.println("</tr>");
}
out.println("</table>");%>
<br>
<hr>
<h3>AddNewBook Details</h3>
<form name="addbook">
<tableborder=1>
<tr> <td>BookTitle</td> <td> <inputtype="text"name="title"> </td>
<tr> <td>AuthorName</td> <td> <input type="text"name="author">
<td>
```

```
    <tr> <td> Publisher</td> <td> <input type="text" name="pub"> </td>
    <tr> <td>Edition</td> <td> <input type="text" name="edition"> </td>
    <tr> <td>Price (in Rs.)</td> <td> <inputtype="text"name="price"> </td>
    <tr> <td>QuantityAvailable(Nos.)</td> <td> <inputtype="text"
    name="qty"> </td>
    <tr> <td>Description</td><td><input type="text" name="desc"> </td>
    <tr> <td>Category</td> <td> <select size="1" name="cat">
    <%
        while(rs1.next())
        {%>
        <optionvalue="<%=rs1.getString(1)%>"><%=rs1.getString(2)%></
    option>
        <%}%>
        </select></td>
        </tr>
        <tr><td><input type="submit" value="Add"onClick="check();"></td>
    <td><inputtype="Reset"value="Clear"></td>
    </table>
        </form>
<%
}
catch(Exceptione)
    {
    out.print("Error="+e+"<HR>");
    }
    %>
</body>
```

```
</html>
```

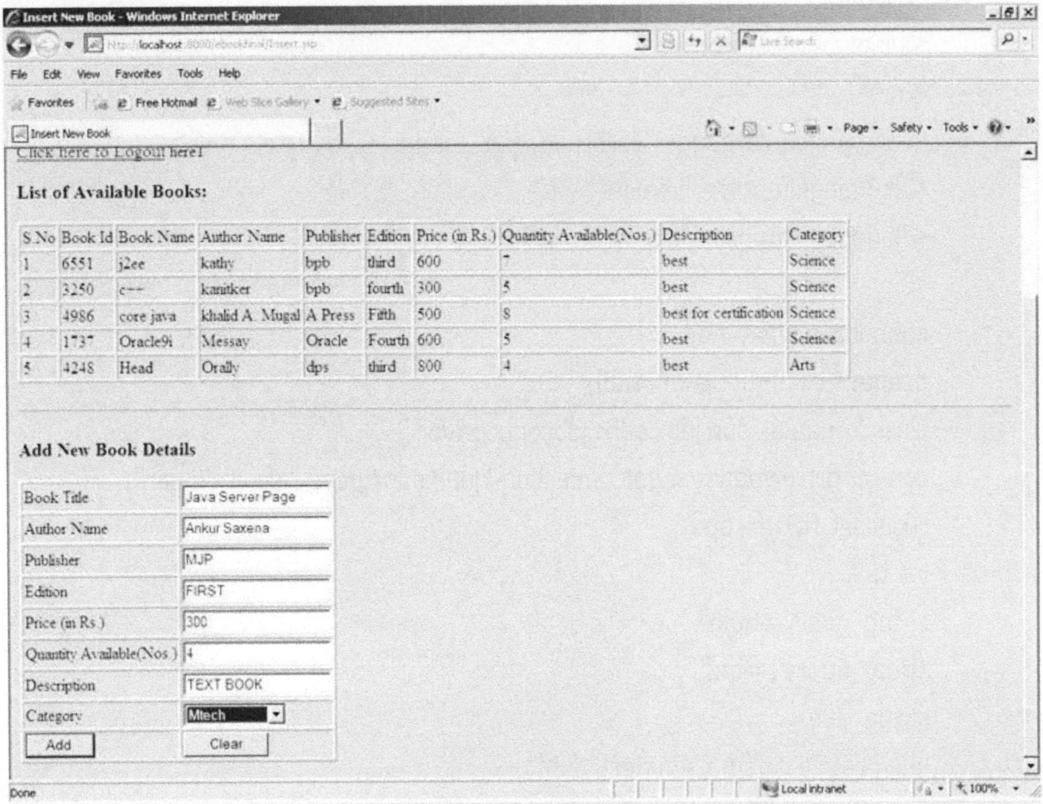

Figure 15.12 Output of Insert.jsp

This screen (Figure 15.12) display the list of available books and add new book details added in this project.

AddBook.jsp

```
<html>
<head>
<title>insert new book</title>
</head>
<bodybgcolor="#ccffff">
    <center>
    <h1>administeringonlineshoppingcart</h1>
```

```
<h2>add book</h2>
</center>
<br>
<hr>
<%@pageerrorpage="errorpage.jsp" import="java.net.*" %>
<%@pageimport="java.io.*" %>
<%@pageimport="java.sql.*" %>
<%
connectioncon= null;
preparedstatementstat=null;
class.forname("sun.jdbc.odbc.jdbcodbcdriver");
con = drivermanager.getconnection("jdbc:odbc:data","scott","tiger");
resultset rs1 = null;
int rs;
string query=null;
string query1=null;
string title;
title=request.getparameter("title");
query1="select a.* from book_details a where a.title=?";
preparedstatement stat1= con.preparestatement(query1);
stat1.setstring(1,title);
rs1=stat1.executequery();
if(rs1.next())
{
%>
<script language="javascript">
alert("this book already exists.");
location.href="insert.jsp";
</script>
<%
}
```

```
else
{
        int int_bookid,int_categoryid,int_price,int_quantity;
        stringstr_title,str_author,str_publisher,str_edition,str_description;
        int_bookid=(int)(10000*math.random()+1);
        int_categoryid=integer.parseint(request.getparameter("cat"));
            str_title=request.getparameter("title");
            str_author=request.getparameter("author");
            str_publisher=request.getparameter("pub");
            str_edition=request.getparameter("edition");
            int_price=integer.parseint(request.getparameter("price"));
            int_quantity=integer.parseint(request.getparameter("qty"));
            str_description=request.getparameter("desc");
            query="insert into book_details values (?,?,?,?,?,?,?,?,?)";
            stat = con.preparestatement(query);
            stat.setint(1,int_bookid);
            stat.setint(2,int_categoryid);
            stat.setstring(3,str_title);
            stat.setstring(4,str_author);
            stat.setstring(5,str_publisher);
            stat.setstring(6,str_edition);
            stat.setint(7,int_price);
            stat.setint(8,int_quantity);
            stat.setstring(9,str_description);
            rs=stat.executeupdate();
            if(rs==1)
            {%>
            <br>
            <h3>successfullyinsertedthedata.</h3>
            <% }
            else
```

```
                {%>
                        <br>
                        <h3>couldnotaddthebook</h3>
            <%}
}%>
            <br>
            <a href="logout.jsp">clickheretologout</a>
            <br> <br>
            <a href="admin.htm">clickheretogotoadministratorpage</a>
            <br><br>
            <a href="insert.jsp">clickheretoaddmorebooks</a>
</body>
</html>
```

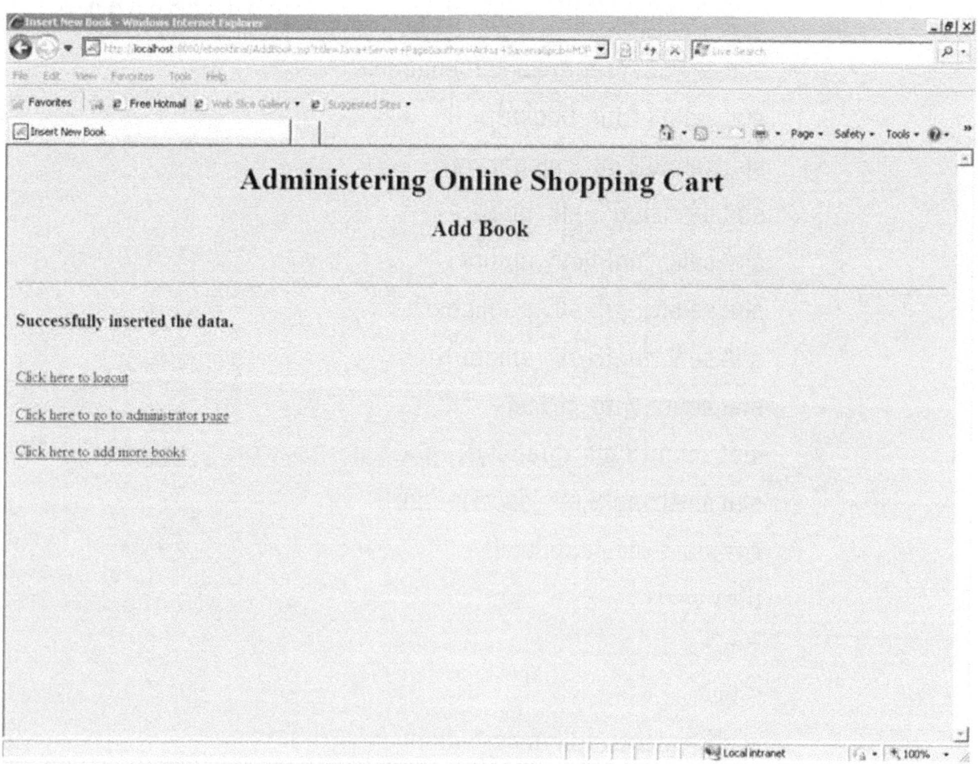

Figure 15.13 Output of Addbook.jsp

This screen (Figure 15.13) display successfully added data in this project.

Insert_Category.jsp

```
<html>
<head>
<title>insertnewcategory</title>
</head>
<bodybgcolor="#ccffff">
    <center>
    <h1>administeringonlineshoppingcart</h1>
    <h2>addcategory</h2>
    </center>
    <br>
    <hr>
    <br>
    <br>
<a href="logout.jsp"><font size="4">clickheretologout</font></a>
<%@pageerrorpage="errorpage.jsp"import="java.net.*" %>
<%@pageimport="java.io.*" %>
<%@pageimport="java.sql.*" %>
<%
    connectioncon=null;
    preparedstatementstat=null;
    resultsetrs=null;
    stringquery=null;
    int ctr=0,flag=0;
    try
    {
    class.forname("sun.jdbc.odbc.jdbcodbcdriver");
    con=drivermanager.getconnection("jdbc:odbc:data","scott","tiger");
    query="select*fromcategory_details";
```

```
        stat=con.preparestatement(query);
        rs=stat.executequery();
        %>
        <br> <h3>addnewcategory:</h3>
        <formname="addbook"action="addcategory.jsp">
        <tableborder=1>
        <tr>
        <td>categoriesavailable</td>
        <td> <selectsize="1"name="cat">
              <%
              while(rs.next())
              {%>
              <option value="<%=rs.getstring(1)%>"> <%=rs.getstring(2)%> </
        option>
                    <%}%>
                    </select> </td>
        </tr>
        <tr>
        <td>categorytobeadded</td>
        <td> <inputtype="text"name="cat_name"> </td> </tr>
        <tr> <td> <inputtype="submit"value="add"> </td><td> <input
        type="reset"value="clear"> </td>
        </table> </form>
<%
        }
        catch(exception e)
        {
        out.print("error = " + e + "<hr>");
        }
        finally
        {
```

```
            rs.close();
            con.close();
            }
        %>
</body>
</html>
addcategory.jsp
<html>
<head>
<title>insert new category</title>
</head>
<bodybgcolor="#ccffff">
        <center>
        <h1>administeringonlineshoppingcart</h1>
        <h2>addcategory</h2>
        </center>
        <br>
        <hr>
        <br>
<%@pageerrorpage="errorpage.jsp"import="java.net.*" %>
<%@pageimport="java.io.*"%>
<%@pageimport="java.sql.*" %>
<%
        connectioncon=null;
        preparedstatementstat=null;
        .forname("sun.jdbc.odbc.jdbcodbcdriver");
        con=drivermanager.getconnection("jdbc:odbc:data","scott","tiger");
        intrs;
        intint_categoryid;
        stringquery=null;
        stringstr_category;
```

```
        int_categoryid=(int)(10000*math.random()+1);
        str_category=request.getparameter("cat_name");
        query="insertintocategory_detailsvalues (?,?)";
        stat=con.preparestatement(query);
        stat.setint(1,int_categoryid);
        stat.setstring(2,str_category);
        rs=stat.executeupdate();
        if (rs==1)
        {
            %> <h3>newcategorysuccessfullyadded</h3> <%
        }
        else
        {
            %> <h3>transactionnotallowed</h3> <%
}
%>
<br>
<ahref="logout.jsp"> <fontsize="4">clickheretologout</font> </a>
<br> <br>
<ahref="admin.htm"> <fontsize="4">clickheretogotoadministratorpage</font>
    < a>
<br> <br>
<a href="insert_category.jsp"> <font size="4">clickheretoaddmorecategories</
    font> </a>
    <br> <br>
</body>
</html>
```

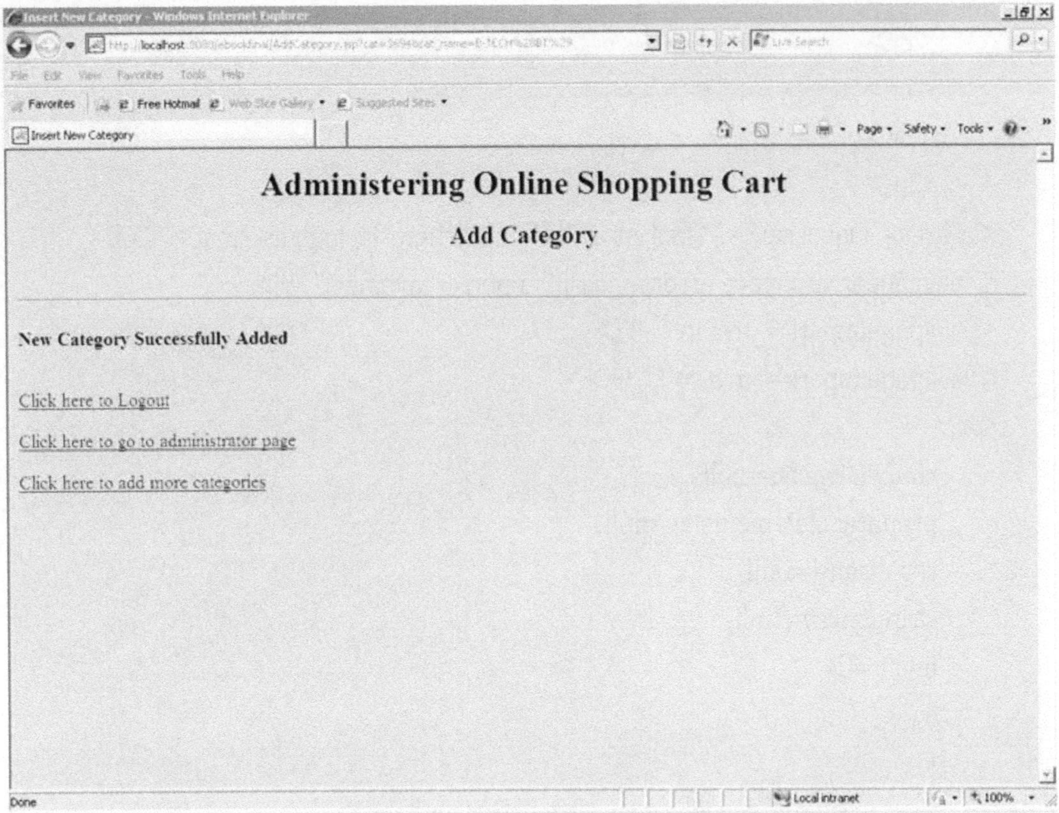

Figure 15.14 Output of Add Catagory.jsp

This screen (Figure 15.14) display successfully added category of the book.

Modify.jsp

```
<html>
<head>
<title>modifybookdetails</title>
</head>
<body bgcolor="#ccffff">
    <center>
    <h1>administeringonlineshoppingcart</h1>
    <h2>modify book details</h2>
```

```
        </center>
        <br>
        <hr>
        <br>
        <br>
    <a href="logout.jsp"> <font size="4">click here to logout</font> </a>
    <%@pageerrorpage="errorpage.jsp"import="java.net.*" %>
    <%@pageimport="java.io.*" %>
    <%@pageimport="java.sql.*" %>
    <%
        connectioncon=null;
        preparedstatementstat=null;
        resultsetrs=null;
        stringquery=null;
        intctr=0;
        try
        {
        class.forname("sun.jdbc.odbc.jdbcodbcdriver");
        con=drivermanager.getconnection("jdbc:odbc:data","scott","tiger");
        query="select a.*,b.categorynamefrombook_details a,category_detailsbwhere
        a.categoryid=b.categoryid";
        stat=con.preparestatement(query);
        rs=stat.executequery();
    %>
        <br> <h3>listofavailablebooks:</h3>
        <formname="mod"action="modify_book.jsp">
        <tableborder=1>
        <tr>
        <td>s.no</td>
        <td>book id</td>
        <td>bookname</td>
```

```
<td>authorname</td>
<td>publisher</td>
<td>edition</td>
<td>price(in rs.)</td>
<td>quantityavailable(nos.)</td>
<td>description</td>
<td>category</td>
</tr>
<%
int int_bookid,int_categoryid,int_price,int_quantity;
string str_title,str_author,str_publisher,str_edition,str_description;
while(rs.next())
{
        ctr=ctr+1;
        out.println("<tr><td>"+ctr);
        int_bookid=rs.getint(1);
        out.println("<td>");
        out.println("<input type=radioname=r1value="+int_bookid+">");
int_categoryid=rs.getint(2);
str_title=rs.getstring(3);
out.println("<td>");
out.println(str_title);
str_author=rs.getstring(4);
out.println("<td>");
out.println(str_author);
str_publisher=rs.getstring(5);
out.println("<td>");
out.println(str_publisher);
str_edition=rs.getstring(6);
out.println("<td>");
out.println(str_edition);
```

```
        out.println("<td>");
        out.println(int_price);
        int_quantity=rs.getint(8);
        out.println("<td>");
        out.println(int_quantity);
        str_description=rs.getstring(9);
        out.println("<td>");
        out.println(str_description);
        out.println("<td>");
        out.println(rs.getstring(10));
        out.println("</tr>");
    }
    %>
        <tr>
            <td> <inputtype="submit"value="modify"> </td>
            <td> <inputtype="reset"value="clear"> </td>
        </tr>
        </table>
        </form>
    <%
        }
        catch(exception e)
        {
        out.print("error = "+e+"<hr>");
        }
        %>
    </body>
    </html>
```

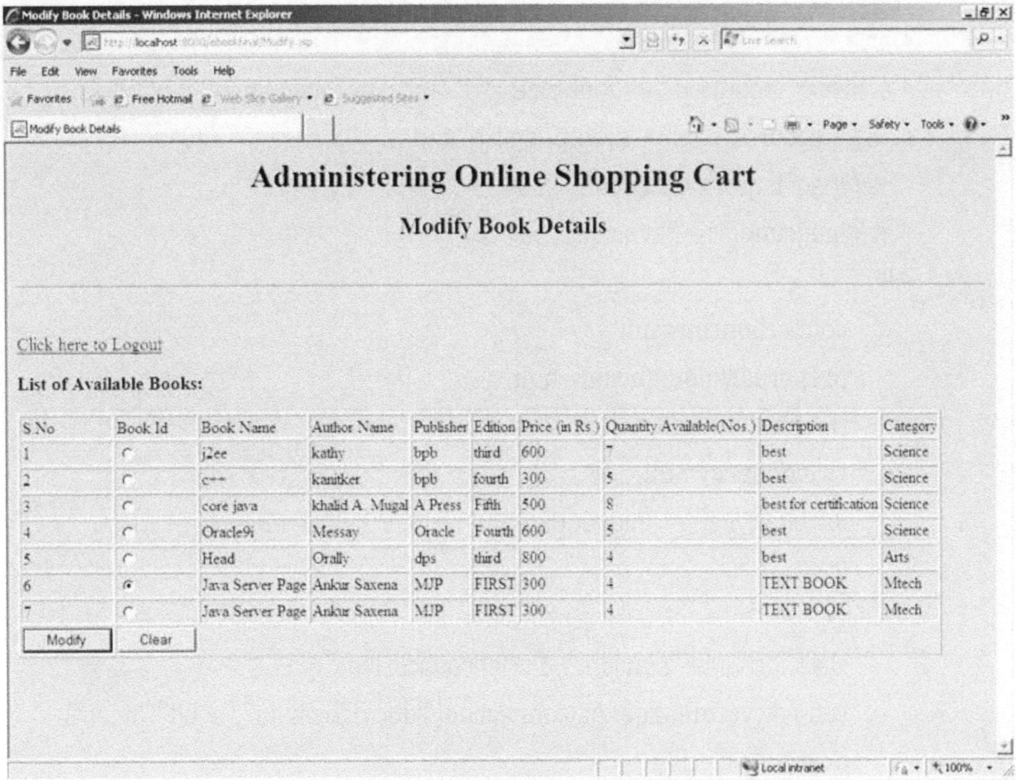

Figure 15.15 Output of Modify.jsp

This screen (Figure 15.15) display to modification of available book.

Modify_Book.jsp

```
<html>
<head>
<title>modifybook</title>
</head>
<bodybgcolor="#ccffff">
    <center>
    <h1>administeringonlineshoppingcart</h1>
    <h2>modifybookdetails</h2>
    </center>
    <hr>
```

```
<br>
<br>
<a href="logout.jsp"><font size="4">clickheretologout</font></a>
<%@pageerrorpage="errorpage.jsp"import="java.net.*" %>
<%@pageimport="java.io.*" %>
<%@pageimport="java.sql.*" %>
<%
    connectioncon=null;
    preparedstatementstat=null;
    resultsetrs=null;
    stringquery=null;
    int book_sel = integer.parseint(request.getparameter("r1"));
    try
    {
    class.forname("sun.jdbc.odbc.jdbcodbcdriver");
    con=drivermanager.getconnection("jdbc:odbc:data","scott","tiger");
    query="select
a.*,b.categorynamefrombook_detailsa,category_detailsbwhere a.bookid=?
anda.categoryid=b.categoryid";
    stat=con.preparestatement(query);
    stat.setint(1,book_sel);
    rs=stat.executequery();
    if(rs.next())
    {
%>
<br> <h3>detailsofselectedbook:</h3>
<formname="mod"action="final_mod_book.jsp">
<inputtype="hidden"name="categoryid"value="<%=rs.getstring(2)%>">
<table border=1>
<tr>
<td>book id</td> <td> <input type="hidden" name="bookid"
```

```
value="<%=book_sel%>"> <%=book_sel%> </td> </tr>
<tr>
<td>bookname</td> <td> <inputtype="text"name="btitle"
value=<%=rs.getstring(3)%>></td>
</tr> <tr>
<td>authorname</td> <td> <inputtype="text"name="bauthor"
value=<%=rs.getstring(4)%>> </td>
</tr> <tr>
<td>publisher</td> <td> <inputtype="text"name="bpub"
value=<%=rs.getstring(5)%>> </td>
</tr> <tr>
<td>edition</td> <td> <inputtype="text"name="bedition"
value=<%=rs.getstring(6)%>> </td>
</tr> <tr>
<td>price(in rs.)</td> <td> <inputtype="text"name="bprice"
value=<%=rs.getstring(7)%>> </td>
</tr> <tr>
<td>quantityavailable(nos.)</td> <td> <inputtype="text"name="bqty"
value=<%=rs.getstring(8)%>> </td>
</tr> <tr>
<td>description</td> <td> <inputtype="text"name="bdesc"
value=<%=rs.getstring(9)%>> </td>
</tr> <tr>
<td>category</td> <td> <input type="text"name="bcat"
value=<%=rs.getstring(10)%>> </td>
</tr> <tr>
<tr> <td> <input type="submit" value=" savechanges "> </td><td> <input
type="reset"value="clear"> </td> </tr>
</table>
</form>
<%
}
```

```
}
    catch(exceptione)
    {
        out.print("error="+e+"<hr>");
    }
    %>
</body>
</html>
```

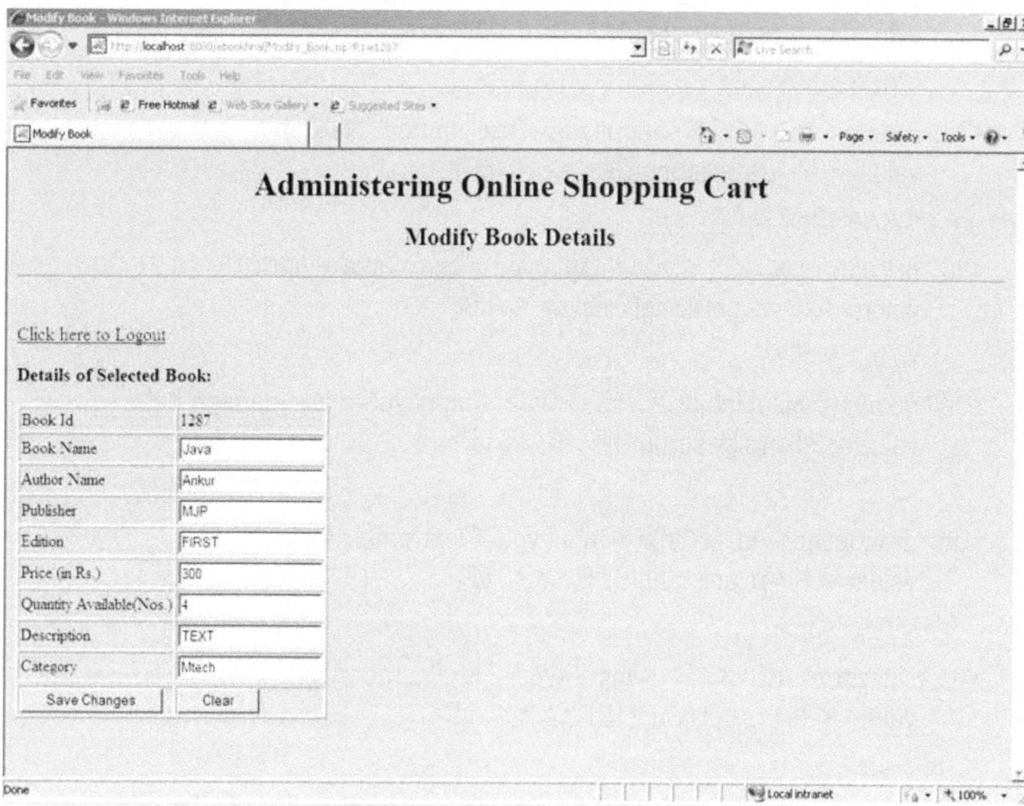

Figure 15.16 Output of Modify_Book.jsp

This screen (Figure 15.16) display modification form of the book.

Final_Mod_Book.jsp

```
<html>
<head>
```

```
<title>bookdetailsmodified</title>
</head>
<bodybgcolor="#ccffff">
    <center>
    <h1> administeringonlineshoppingcart</h1>
    <h2>bookdetailsmodified</h2>
    </center>
    <br>
    <hr>
    <br>
    <br>
    <%@pageerrorpage="errorpage.jsp"import="java.net.*" %>
    <%@pageimport="java.io.*" %>
    <%@pageimport="java.sql.*" %>
    <%
        connectioncon=null;
        //statementstat=null;
        preparedstatementstat=null;
        //statementstat1=null;
        preparedstatementstat1=null;
        resultsetrs=null;
        class.forname("sun.jdbc.odbc.jdbcodbcdriver");
        con=drivermanager.getconnection("jdbc:odbc:data","scott","tiger");
        intrs_int;
        intint_bookid,int_categoryid,int_price,int_quantity;
stringquery=null;
stringstr_title,str_author,str_publisher,str_edition,str_description;
    int_bookid=integer.parseint(request.getparameter("bookid"));
    int_categoryid=integer.parseint(request.getparameter("categoryid"));
    str_title=request.getparameter("btitle");
    str_author=request.getparameter("bauthor");
```

```
        str_publisher=request.getparameter("bpub");
        str_edition=request.getparameter("bedition");
        int_price=integer.parseint(request.getparameter("bprice"));
        int_quantity=integer.parseint(request.getparameter("bqty"));
        str_description=request.getparameter("bdesc");
    query="updatebook_details set
        categoryid=?,title=?,author=?,publisher=?,edition=?,price=?,quantity=?,description=?
        where bookid=?";
        stat = con.preparestatement(query);
        stat.setint(1,int_categoryid);
        stat.setstring(2,str_title);
        stat.setstring(3,str_author);
        stat.setstring(4,str_publisher);
        stat.setstring(5,str_edition);
        stat.setint(6,int_price);
        stat.setint(7,int_quantity);
        stat.setstring(8,str_description);
        stat.setint(9,int_bookid);
        rs_int=stat.executeupdate();
        if (rs_int==1)
        {
            out.println("bookdetailsforbookid "+int_bookid+"successfullymodified.");
        }
        else
        {
            out.println("sorrytransactionnotcompleted");
        }
%>
<br> <br>
<a href="logout.jsp">clickheretologout</a>
<br> <br>
```

```
<a href="admin.htm">clickheretogotoadministratorpage</a>
<br> <br>
<a href="modify.jsp">clickheretomodifymorebooks</a>
</body>
</html>
```

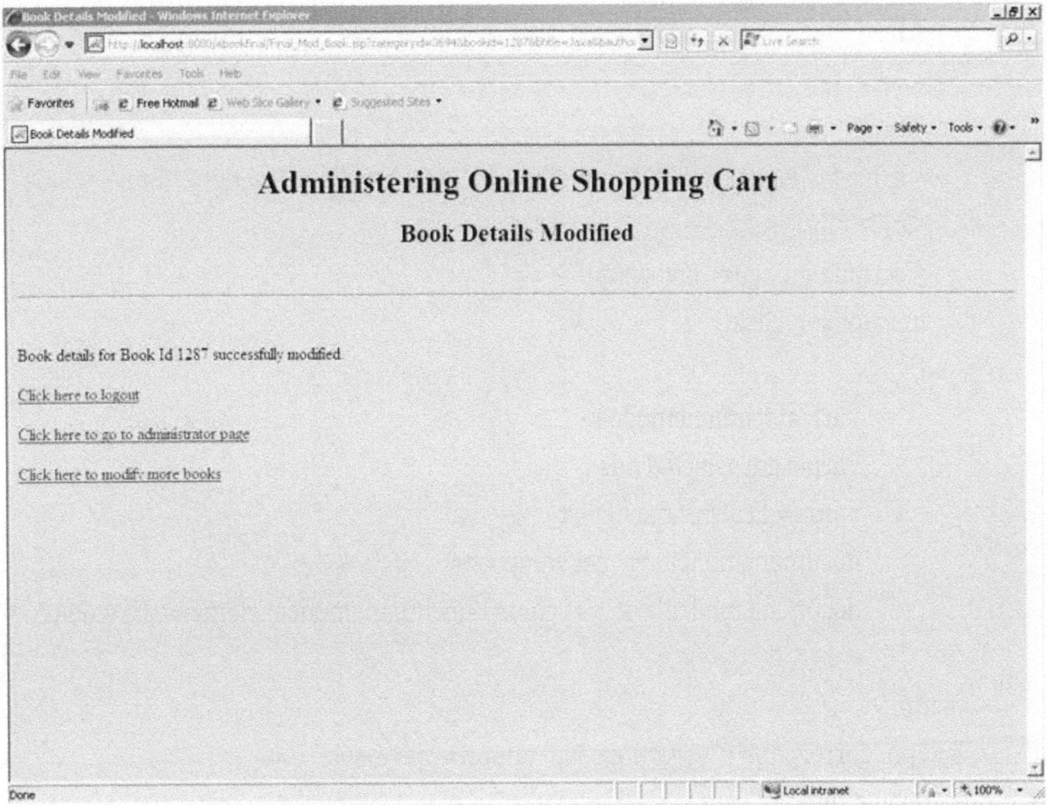

Figure 15.17 Output of Final_Mod_Book.jsp

This screen (Figure 15.17) display successfully, modified information of the book.

Update_Category.jsp

```
<html>
<head>
<title>categorymodified</title>
```

```
</head>
<bodybgcolor="#ccffff">
    <center>
    <h1> administeringonlineshoppingcart</h1>
    <h2>categorymodified</h2>
    </center>
    <br>
    <hr>
    <br>
    <a href="logout.jsp"> <font size="4">click here to logout</font> </a>
    <br> <br>
    <scriptlanguage="javascript">
    functionsel_click()
    {
        varf=document.modcat;
        vari=f.cat.selectedindex;
        varstr=f.cat.options[i].text;
        document.modcat.vr_cat.value=str;
        document.modcat.cat_val.value=document.modcat.elements[0].value;
}
</script>
<%@pageerrorpage="errorpage.jsp"import="java.net.*" %>
<%@pageimport="java.io.*" %>
<%@pageimport="java.sql.*" %>
< %
    connectioncon=null;
    preparedstatementstat=null;
    resultsetrs=null;
    stringquery=null;
    try
    {
```

```
        class.forname("sun.jdbc.odbc.jdbcodbcdriver");
        con=drivermanager.getconnection("jdbc:odbc:data","scott","tiger");
        query="select*fromcategory_details";
        stat=con.preparestatement(query);
        rs=stat.executequery();
    %>
        <br> <h2>modifycategory:--</h2>
        <formname="modcat"action="modifycategory.jsp">
        <tableborder=1>
        <tr>
        <td>category</td>
        <td> <select size="1"name="cat"onchange="sel_click();">
            <%
            while(rs.next())
            {%>
            <optionvalue="<%=rs.getstring(1)%>">
            <%=rs.getstring(2)%> </option>
            <%}%>
            </select> </td>
     </tr>
     <tr><tdcolspan=2> <inputtype="text"name="vr_cat"> <inputtype="text"
         name="cat_val"> </td>
     <tr> <td> <inputtype="submit"value="modify"> </td> <td> <input
         type="reset"value="clear"></td>
         </table> </form>
     <%
         }
         catch(exception e)
         {
         //out.print("error="+e+"<hr>");
         }
```

```
    finally
    {
        rs.close();
        con.close();
    }
    %>
</body>
</html>
```

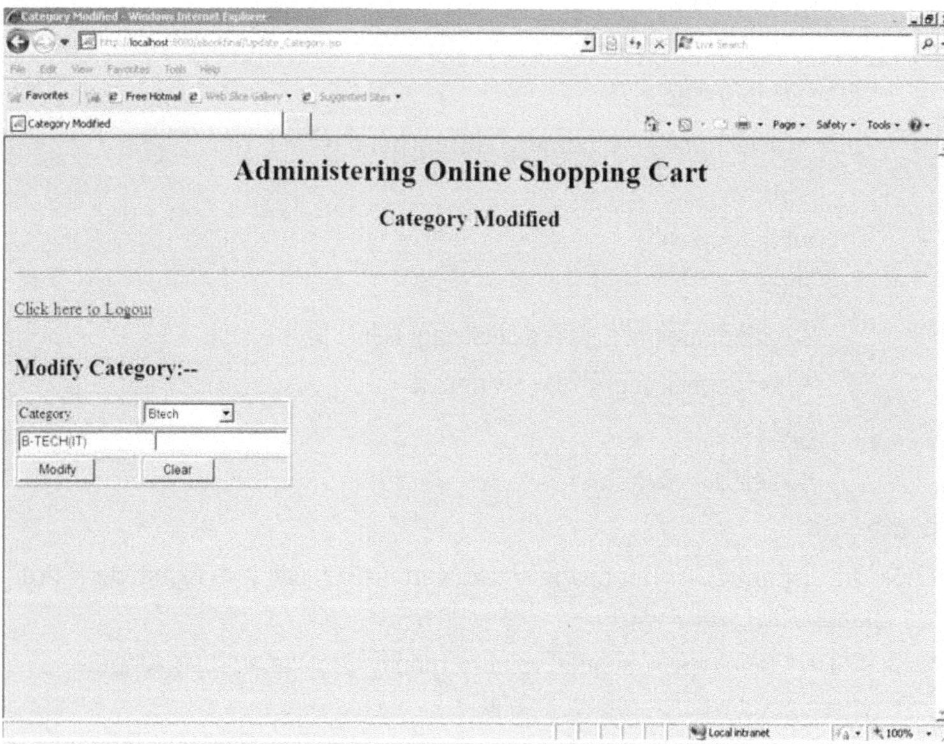

Figure 15.18 Output of Update_Catagory.jsp

This screen (Figure 15.18) display form for category modified of the book.

ModifyCategory.jsp

```
<html>
<head>
<title>modifycategory</title>
```

```
</head>
<bodybgcolor="#ccffff">
    <center>
    <h1> administeringonlineshoppingcart</h1>
    </center>
    <br>
    <hr>
    <br>
    <br>
<%@pageerrorpage="errorpage.jsp"import="java.net.*" %>
<%@pageimport="java.io.*" %>
<%@pageimport="java.sql.*" %>
<%
    connectioncon=null;
    preparedstatementstat=null;
    intrs=0;
    stringquery=null;
    try
    {
    stringnew_cat="";
    intcat_id=0;
    new_cat=request.getparameter("vr_cat");
    cat_id=integer.parseint(request.getparameter("cat_val"));
    class.forname("sun.jdbc.odbc.jdbcodbcdriver");
    con=drivermanager.getconnection("jdbc:odbc:data","scott","tiger");
    query="update category_details set categoryname=? where categoryid=?";
    stat=con.preparestatement(query);
    stat.setstring(1,new_cat);
    stat.setint(2,cat_id);
    rs=stat.executeupdate();
    if (rs==1)
```

```
{
     out.println("categorysuccessfullymodified");
}
else
{
     out.println("transactionnotallowed");
}
}
catch(exception e)
{
     out.print("error="+e+"<hr>");
}
finally
{
     con.close();
}
%>
     <br>
     <br>
     <a href="logout.jsp">clickheretologout</a>
     <br> <br>
     <a href="admin.htm">clickheretogotoadministratorpage</a>
     <br> <br>
     <a href="update_category.jsp">clickheretomodifymorecategories</a>
</body>
</html>
```

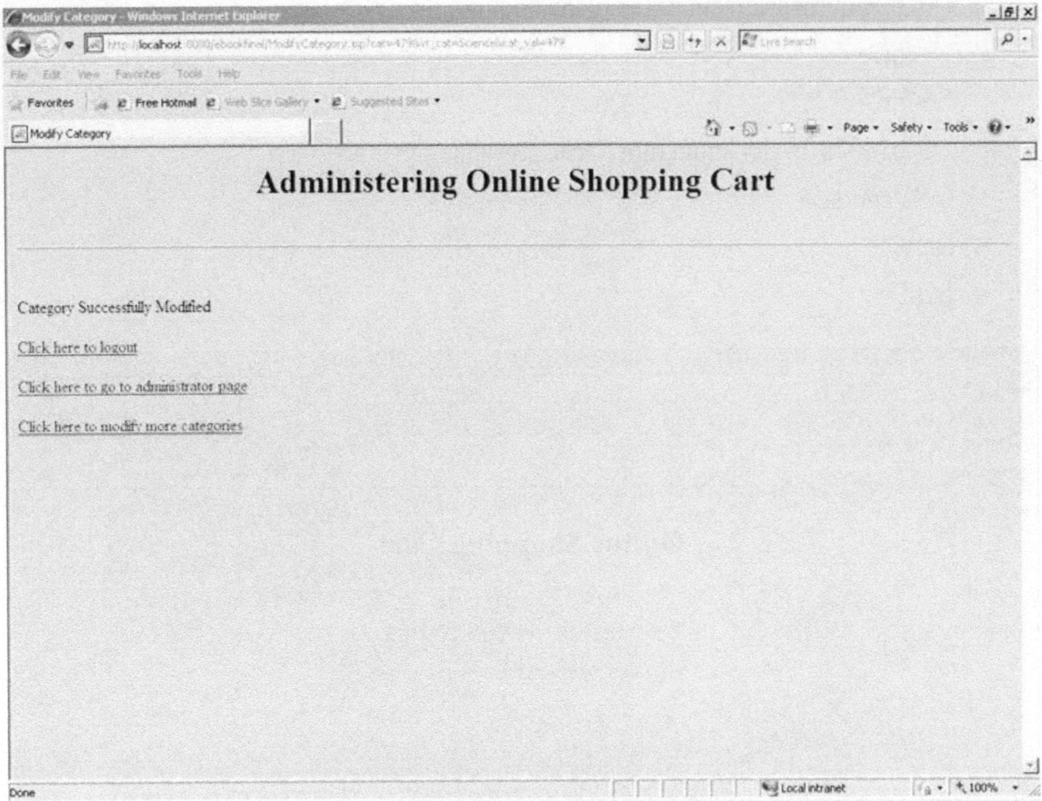

Figure 15.19 Output of Modify_Catagory.jsp

This screen (Figure 15.19) display successfully added category of the book.

Unauthorised.htm

```
<html>
<head>
<title>invaliduser</title>
</head>
<bodybgcolor="#ccffff">
    <br>
    <center>
    <h1>onlineshoppingcart</h1>
    <br>
    <br>
```

```
        <b>yoursessionisnotvalid.pleaseloginagain.</b>
        <br>
        <br>
        <b> <a href="login.htm">clickheretologin</a> </b>
        </center>
    </body>
    </html>
```

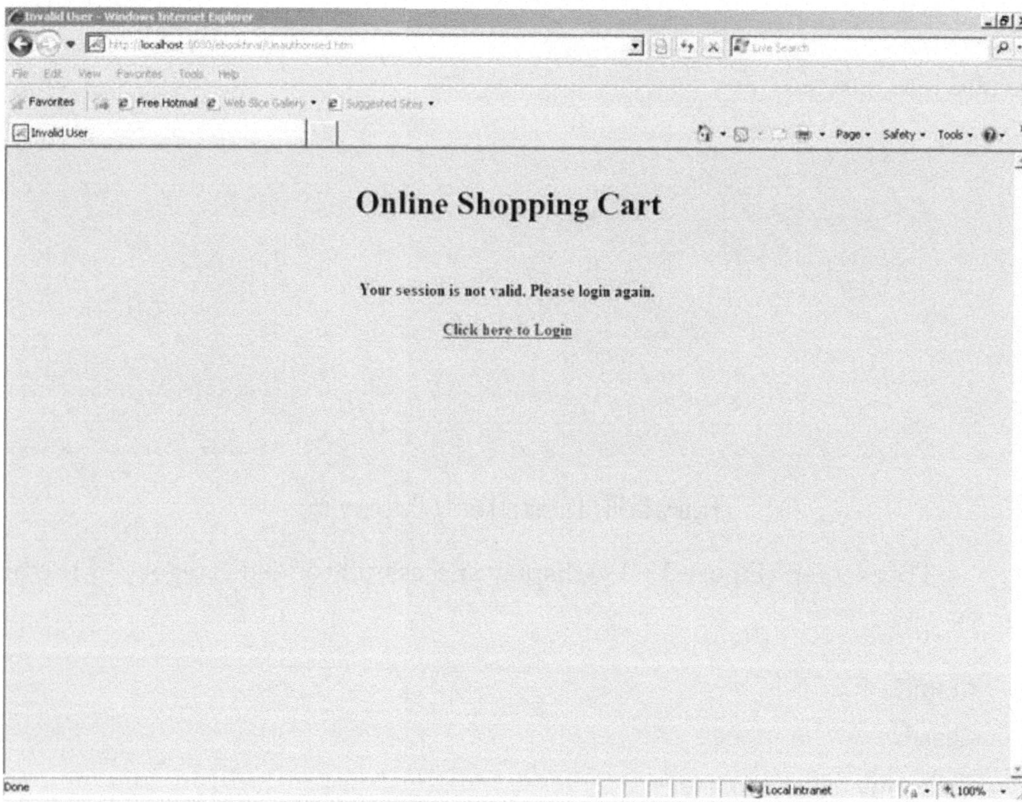

Figure 15.20 Output of Unauthorised.htm

This screen (Figure 15.20) display session information of the project.

Errorpage.jsp

```
<%@pageisErrorPage="true"import="java.io.PrintWriter"%>
<html>
<head>
```

```
<title>
    ShoppingCartErrorPage
</title>
</head>
<body>
    <h2>ErrorPage</h2>
    <br>
    <h4>Anerrorhasocurredintheapplication</h4>
    <br> <p>
    Thestacktraceoftheerroris:
    < p >
    <%=exception.toString() %>
    <%--exception.printStackTrace(newPrintWriter(out)); --%>
</body>
</html>
```

Inter_Cart.jsp

```
<%@pageerrorPage="errorpage.jsp"language="java" %>
< %
    int i;
    Stringbookid=null;
    Stringuser_src=null;
    user_src=(String)session.getValue("user");
    if(user_src!=null)
    {
        int counter=0;
        counter=Integer.parseInt((String)session.getValue("ctr_val"));
        int ctr=0;
        for(i=1;i<=counter;i++)
        {
            if(request.getParameter("chk"+i)!=null)
            {
```

```
                bookid=request.getParameter("chk"+i);
                ctr=ctr+1;
                session.putValue("chk_var"+ctr,bookid);
            }
        }
        session.putValue("TotalSel",String.valueOf(ctr));
        response.sendRedirect("Cart.jsp");
    }
    else
    {
        response.sendRedirect("Unauthorised.htm");
    }
    % >
```

NewSearch.jsp

```
<html>
<head>
<title>searchresults</title>
</head>
<bodybgcolor="#ccffff">
    <center>
    <h1> <u >onlineshoppingcart</u > </h1>
    </center>
    <br>
    <a href="logout.jsp"> <font size="4">clickheretologout</font> </a>
    <%@pageerrorpage="errorpage.jsp"language="java"import="java.sql.*"%>
<%
stringuser_src=(string)session.getvalue("user");
if(user_src!=null)
{
stringstr_searchvar=null;
```

```
stringstr_colvar=null;
stringquery="";
stringtab_var="";
inttotal_sel=integer.parseint((string)session.getvalue("totalsel"));
int[]old_bookid=newint[total_sel];
connectionconn;
conn=null;
resultsetrs;
rs=null;
class.forname("sun.jdbc.odbc.jdbcodbcdriver");
conn=drivermanager.getconnection("jdbc:odbc:data","scott","tiger");
preparedstatement stat=null,stat_ins=null,stat_del=null;
for (int i=1;i<=total_sel;i++)
{
old_bookid[i-1]=integer.parseint((string)session.getvalue("chk_var"+i));
}%>
    <h3>yourcartdetailsareasfollows:</h3>
    <formname="shop"action="cart_quantity.jsp">
    <tableborder=1>
    <tr>
    <td>s.no</td>
    <td>bookname</td>
    <td>price(in rs.)</td>
    </tr>
    <%
    intprice=0; intbookprice=0;
    for(int j=0;j<total_sel;j++)
    {
        out.println("<tr> <td>"+j+1);
        query="select title, pricefrombook_detailswherebookid=?";
        stat=conn.preparestatement(query);
```

```
            stat.setint(1,old_bookid[j]);
            rs=stat.executequery();
            while(rs.next())
            {
                out.println("<td>");
                out.println(rs.getstring(1));
                out.println("<td>");
                price=rs.getint(2);
                out.println(price);
        }
        rs.close();
        bookprice=bookprice+price;
        out.println("</tr>");
}%>
<tr>
<tdcolspan=3>totalcartamount(in rs.)=
<%=bookprice%>
<td> </tr>
</table>
<br>
<inputtype=submitvalue="placeorder">
</form>
<%
stringtest_src=null;
intchk_ctr=0;
if(request.getparameter("r1").equals("title"))
{
    str_searchvar=request.getparameter("booktitle");
    str_colvar="title";
}
if(request.getparameter("r1").equals("category"))
```

```
{
    chk_ctr=1;
    str_searchvar=request.getparameter("bookcategory");
    str_colvar="categoryname";
}
if(request.getparameter("r1").equals("publisher"))
{
    str_searchvar=request.getparameter("bookpublisher");
    str_colvar="publisher";
}
if(request.getparameter("r1").equals("author"))
{
    str_searchvar=request.getparameter("bookauthor");
    str_colvar="author";
}
stringss=request.getparameter("r2");
if(request.getparameter("r2")==null)
{
    tab_var="book_details";
}
else
{
// togetthenameofthetableinwhichthesearchtobedone-------
    if(request.getparameter("r2").equals("a"))
    {
        tab_var="temp_detail";
}
else
{
    tab_var="book_details";
}
```

```
}
//-------- whethersearchisbycategoryorothercriteria-----------
if (chk_ctr==0)
{
    query="selecta.*from "+tab_var+"a,category_detailsbwhere a."+str_colvar+"
    = ? and a.categoryid=b.categoryid";
}
else
{
    query="select a.*,b.categoryname from "+tab_var+"a,category_details b where
    b."+str_colvar+"=? and a.categoryid=b.categoryid";
}
try
{
    stat=conn.preparestatement(query);
    stat.setstring(1,str_searchvar);
    rs = stat.executequery();
    int ctr=0;%>
    <h3>searchresultsareasfollows:</h3>
    <formname="cart"action="newinter_cart.jsp" method="post">
    <tableborder=1>
    <tr>
    <td>s.no</td>
    <td>select</td>
    <td>bookname</td>
    <td>authorname</td>
    <td>publisher</td>
    <td>edition</td>
    <td>price(in rs.)</td>
    <td>quantityavailable(nos.)</td>
    <td>description</td>
```

```
</tr>
<%
stringdelete_query="deletefromtemp_detail";
stat_del= conn.preparestatement(delete_query);
stringins_query=null;
intint_bookid,int_categoryid,int_price,int_quantity;
stringstr_title,str_author,str_publisher,str_edition,str_description;
intrs_del=stat_del.executeupdate();
while(rs.next())
{
        ctr=ctr+1;
        out.println("<tr> <td>"+ctr);
        int_bookid=rs.getint(1);
out.println("<td> <input type=checkboxname=chk"+ctr+"
value="+int_bookid+">");
        int_categoryid=rs.getint(2);
        str_title=rs.getstring(3);
out.println("<td>");
out.println(str_title);
str_author=rs.getstring(4);
out.println("<td>");
out.println(str_author);
str_publisher=rs.getstring(5);
out.println("<td>");
out.println(str_publisher);
str_edition=rs.getstring(6);
out.println("<td>");
out.println(str_edition);
int_price=rs.getint(7);
out.println("<td>");
out.println(int_price);
```

```
        int_quantity=rs.getint(8);
        out.println("<td>");
        out.println(int_quantity);
        str_description=rs.getstring(9);
        out.println("<td>");
        out.println(str_description);
        out.println("</tr>");
        ins_query="insert into temp_detail values(?,?,?,?,?,?,?,?,?)";
        stat_ins=conn.preparestatement(ins_query);
        stat_ins.setint(1,int_bookid);
        stat_ins.setint(2,int_categoryid);
        stat_ins.setstring(3,str_title);
        stat_ins.setstring(4,str_author);
        stat_ins.setstring(5,str_publisher);
        stat_ins.setstring(6,str_edition);
        stat_ins.setint(7,int_price);
        stat_ins.setint(8,int_quantity);
        stat_ins.setstring(9,str_description);
        int rs_ins=stat_ins.executeupdate();
    }
    string str_ctr=string.valueof(ctr);
    session.putvalue("ctr_val",str_ctr);
    %>
    </table>
    <input type="submit"value="addtocart"name="add">
    </form>
    <%
    resultset rscombo=null;
        string query1="select * from category_details";
        preparedstatement stat_combo = conn.preparestatement(query1);
        rscombo=stat_combo.executequery();%>
```

```
<br> <h3>search results are as follows:</h3>
<form name="newsrc" action="newsearch.jsp" method="post">
<table border=1>
    <tr>
<td> <input type="radio" value="category" checked name="r1"> </td>
<td>search based on category</td>
<td>
    <select size="1" name="bookcategory">
        <option selected value="select">select category</option>
    <%
    while(rscombo.next())
    {
        stringcategory=rscombo.getstring(2);
    %>
        <option value="<%=category%>"> <%=category%> </option>
        <%}%>
    </select>
</td>
</tr>
    <tr>
    <td> <input type="radio" name="r1" value="title"> </td>
    <td>search by book title</td>
    <td> <input type="text" name="booktitle" size="25"> </td>
</tr>
<tr>
<td> <input type="radio" name="r1" value="author"> </td>
    <td>search by book author</td>
    <td> <input type="text" name="bookauthor" size="25"> </td>
</tr>
<tr>
    <td> <input type="radio" name="r1" value="publisher"> </td>
```

```
    <td>search by book publisher</td>
    <td> <input type="text" name="bookpublisher" size="25"> </td>
</tr>
    <tr>
    <td> <input type="radio" name="r2" value="a">advanced search</td>
    <td> <input type="radio" name="r2" value="n" checked>new search</td>
    <td> <input type=submit value="search"> </td>
</tr>
</table>
</form>
<%}
catch(exception e)
{
    out.println("error  "+e);
}
finally{conn.close();}
}
else
    {
        response.sendRedirect("Unauthorised.htm");
    }
%>
</body>
</html>
```

REFERENCES

1. Budi Kurniawan, (2002) *Java for the Web with Servlets, JSP and EJB*, Techmedia.
2. www.roseindia.net
3. www.java.sun.com
4. www.jsptut.com